The Space and Power of Young People's Social Relationships

The book examines the power of young people's social relationships in schools to transform, or more often, to continue, differences that pervade societies: mind-body-emotional differences or Special Educational Needs and Disability, gender, poverty, race/ethnicity, sexuality and their intersections. The book details extensive qualitative research with young people, foregrounding their accounts.

In challenging educators and others to engage with young people's own agencies and to make space for their socialities, the concepts of embodied social and emotional capital and young people as contextual bodies/subjectivities/agencies are developed, emphasising both young people's agencies and how these are socio-spatially situated, constrained and enabled. The book is most concerned with how and when young people challenge and change enduring differences. The concept of 'immersive geographies' outlines the potential of change inherent in the repeated coming together of the same people in space, doing similar things that are, however, always provisional and always with the potential to be done differently. Examples of when difference is transformed are presented.

The book marks a major interdisciplinary contribution to geographies and social studies of children, youth and education, child development, social work, social policy and education studies. Furthermore, it is of appeal to anyone interested in young people, social reproduction and sociality: from educators, policy makers, youth workers and social workers to parents.

Louise Holt is Professor of Human Geography at Loughborough University and an internationally recognised scholar in the fields of geographies and social studies of children and youth and social and cultural geography. Louise was Editor-in-Chief of the journal *Children's Geographies* for many years and founded the International Conferences of Geographies of Children, Youth and Families. She has published three key edited collections and many journal articles.

Routledge Spaces of Childhood and Youth Series

Edited by Peter Kraftl and John Horton

The *Routledge Spaces of Childhood and Youth Series* provides a forum for original, interdisciplinary and cutting edge research to explore the lives of children and young people across the social sciences and humanities. Reflecting contemporary interest in spatial processes and metaphors across several disciplines, titles within the series explore a range of ways in which concepts such as space, place, spatiality, geographical scale, movement/mobilities, networks and flows may be deployed in childhood and youth scholarship. This series provides a forum for new theoretical, empirical and methodological perspectives and ground-breaking research that reflects the wealth of research currently being undertaken. Proposals that are cross-disciplinary, comparative and/or use mixed or creative methods are particularly welcomed, as are proposals that offer critical perspectives on the role of spatial theory in understanding children and young people's lives. The series is aimed at upper-level undergraduates, research students and academics, appealing to geographers as well as the broader social sciences, arts and humanities.

Why Garden in Schools?
Lexi Earl and Pat Thomson

Latin American Transnational Children and Youth
Experiences of Nature and Place, Culture and Care Across the Americas
Edited by Victoria Derr and Yolanda Corona-Caraveo

Mapping the Moral Geographies of Education
Character, Citizenship and Values
Sarah Mills

The Space and Power of Young People's Social Relationships
Immersive Geographies
Louise Holt

For more information about this series, please visit: https://www.routledge.com/Routledge-Spaces-of-Childhood-and-Youth-Series/book-series/RSCYS

The Space and Power of Young People's Social Relationships

Immersive Geographies

Louise Holt

Routledge
Taylor & Francis Group
LONDON AND NEW YORK

First published 2024
by Routledge
4 Park Square, Milton Park, Abingdon, Oxon OX14 4RN

and by Routledge
605 Third Avenue, New York, NY 10158

Routledge is an imprint of the Taylor & Francis Group, an informa business

British Library Cataloguing-in-Publication Data
A catalogue record for this book is available from the British Library

Library of Congress Cataloging-in-Publication Data
Names: Holt, Louise, 1975– author.
Title: The space and power of young people's social relationships : geographies of immersion / Louise Holt.
Description: Abingdon, Oxon ; New York, NY : Routledge, 2024. | Series: Routledge spaces of childhood and youth | Includes bibliographical references and index.
Identifiers: LCCN 2023027401 (print) | LCCN 2023027402 (ebook) | ISBN 9780367463236 (hardback) | ISBN 9781032612621 (paperback) | ISBN 9781003028161 (ebook)
Subjects: LCSH: Socialization. | Interpersonal relations in children. | Human geography.
Classification: LCC HQ783 .H656 2024 (print) | LCC HQ783 (ebook) | DDC 155.4/192—dc23/eng/20231108
LC record available at https://lccn.loc.gov/2023027401

ISBN: 978-0-367-46323-6 (hbk)
ISBN: 978-1-032-61262-1 (pbk)
ISBN: 978-1-003-02816-1 (ebk)

DOI: 10.4324/9781003028161

Typeset in Times New Roman
by Apex CoVantage, LLC

This book is dedicated to my family.

In special loving memory of Sharon, Robert and Paul.

Contents

List of figures — *viii*
Acknowledgements — *ix*
List of abbreviations — *xi*

1 The spaces and power of young people's friendships: immersive geographies — 1

2 Methods, approaches, contexts — 23

3 Young people's friendships and power — 35

4 Immersive geographies — 56

5 Young people's friendships: embodied, emotional and social capital — 75

6 Young people as nodes of the intergenerational reproduction of enduring differences: (re)producing subjectivities — 86

7 Immersive geographies and imminent transformation: young people's powerful socialities – the power to challenge and change enduring inequalities — 114

8 Constraining and enabling young people's power: reflections on the social-spatial contexts of schools — 126

9 Conclusions — 148

References — *155*
Appendices — *169*
Index — *175*

Figures

0.1	Connections	xiii
0.2	An orchard	xiv
4.1	Children's connected bodies in the playground	63
5.1	The plasticine model made by Mahal and Jasmin	76
6.1	A football	87
6.2	Friends	107
6.3	My pet	108
6.4	My trampoline	109
7.1	The child-wheelchair-child-child-child-child assemblage	117
7.2	The sloping field (photograph Jennifer Lea)	119
8.1	Plan of 'bottom set' numeracy, Church Street	130
8.2	Seating plan 'top set' English Year Five	131

Acknowledgements

Sincere thanks go to all the children, young people and adults who participated in the research. You young people have stayed with me always, and I wonder where you are now. I only hope I did your lively and wonderful selves some justice in the pages of the book, but it is inevitably a limited and 2D representation. Express thanks to Jennifer Lea and Sophie Bowlby, partners in the ESRC project. Particular thanks and acknowledgement to Jennifer, who conducted most of the ESRC-funded research and whose distinctive voice can be witnessed in some of the research diaries. Thanks also to Gudbjorg Ottosdottir, who conducted the research in the urban selective girls high school.

I acknowledge and thank the funders who have supported the various studies that have contributed to the material in the book: Loughborough University; the University of Brighton; the Royal Geographical Society with the Institute of British Geographers Engineering and Physical Sciences Research Council Grant; and, the Economic and Social Research Council (ref ES/F033648/1 and ES/X001431/1).

Special thanks are preserved for my wonderful family for the support you have given me whilst I wrote the book and in everything. To my mum, Teresa, and dad, Grahame, my mother-in-law, Elaine, and Mark, Shirley, Abigail and Rebecca. I have to give particular thanks and love to my wonderful children, Iola, Amelie and Sebastian. You are amazing and provide me with so much love and support, and I try to do the same for you. I am not sure what the typical family is, but I fear we are not it; nevertheless, we are doing great, getting by, and I am proud of you for the wonderful people you are. To Darren, my partner in everything and greatest support and honest critic. Thank you all for your patience when I have been partly in the room and partly absent, thinking about my book and for supporting me always in everything.

Thank you, Mr Fox, for your faith in me and your critical, lively and inspirational take on geography. I was so sorry to hear that you left the world too early, and younger than I am now. One of your many legacies is in this book, as a testament to all inspiring teachers who change the world for the better every day.

Thanks to all the friends and colleagues who have heard and commented upon elements of the book over the years. Special thanks to Gabriela Tebet, who invited me to deliver a keynote speech at the 6th International Conference on Geographies of Children, Youth and Families at UNICAMP, State University of Campinas,

Brazil in 2018, and to all the participants who gave feedback on the paper presented there, which was an early version of this book. Very special thanks are due to Chris Philo, Sarah Holloway, Sarah Mills, James Esson and John Harrison for your critical and insightful comments on a full draft of the book.

Sincere thanks to colleagues at Loughborough University for your support and conversations. Thank you to the Education and Student Experience Team of the School of Social Sciences and Humanities. Rachel Breen, Susan Knight and the Student Administrators and Managers; you are amazing and without your efficiency and kindness I could not have done my job as Associate Dean, let alone finished the book. Thanks also to Nick Clifford, who first suggested I should write a book.

The work of Marian Trudgill for your editing work and Sophie Milnes for doing the glossary is much appreciated. Tina Byrom from Loughborough University Enhanced Academic Practice, thank you for your final proof read of the book. I cannot believe you noticed one letter with the incorrect font. My mum, Teresa Rowe, also proof read the book at an earlier stage and again the latest revision.

Final thanks to Iola Smith for the cover photo and some photos in the book and Amelie and Sebastian Smith for some artwork within the book.

Any mistakes, inaccuracies or things that should not have been committed to print are, of course, my own responsibility.

Abbreviations

ADHD: Attention Deficit and Hyperactivity 'Disorder'. A 'condition' of inattention, hyperactivity and impulsiveness which may be co-existent or independent which impacts everyday activities and education. There is a great deal of controversy surrounding potential over (and under) diagnosis and excessive medication (Kazda et al., 2019).

ASD: Autism Spectrum 'Disorders'. A term deployed by some schools to refer to young people on the Autism Spectrum. My preferred term is difference or just being on the Autism Spectrum or neurodiverse. Taken verbatim from adults' quotes.

EHCP: Education and Health Care Plan. These replaced 'statements' in the 2014 Children and Families Act. The highest level of support for young people identified as experiencing SEND. It set out legal entitlements for support for the young person.

LA: Local Authority. Local authorities vary in size. They are the local area organisation of education and other local services in the UK administering around 200–300 schools. Their educational powers and responsibilities have been reduced in recent decades with a move towards centralisation and devolved management of education to educational 'trusts' – an association of schools. Local authorities retain power and responsibility for administering and managing processes pertaining to Special Educational Needs.

LSA: Learning Support Assistant: An adult who is not usually a qualified teacher who assists in classrooms, often funded by a young person's Education and Health Care Plan.

Ofsted: Office for Standards in Education, Children's Services and Skills: Inspect education of schools and colleges (not standard University courses) and children's care services.

SEMHD: Social Emotional and Mental Health 'Difficulties'. A change in definition if the 2014 Code of Practice includes mental health. Previous definitions focused on behaviour. I prefer a focus on 'differences' to emphasise that these can be troubling mind-body-emotional states for young people, but that the idea of 'difficulty' is also socio-spatially situated.

SEND/SEN: Special Educational Needs and Disabilities or SEN, as defined in Section 20 of the 2014 Children and Families Act: "A child or young person has special educational needs if he or she has a learning difficulty or disability which calls for special educational provision to be made for him or her. A child or a young person of compulsory school age has a learning difficulty or disability if he or she has a significantly greater difficulty in learning than the majority of others of the same age, or has a disability which prevents or hinders him or her from making use of facilities of a kind generally provided for others of the same age in mainstream schools or mainstream post-16 institutions". Previous definitions of Special Educational Needs excluded disability.

SENCO/SENDCO: Special Educational Needs and Disabilities Co-ordinator: A specialist teacher who oversees the provision of learning support and ensures the effective education of students with Special Educational Needs and Disabilities in schools or academy trusts.

SATs: Standard Assessment Tests: SATs are standardised assessment tests[1] taken by primary school-aged children at approximately age 7 and age 11. The results of age 11 tests are published and used as an indicator of the success of schools.

Figure 0.1 Connections

Source: Iola Smith

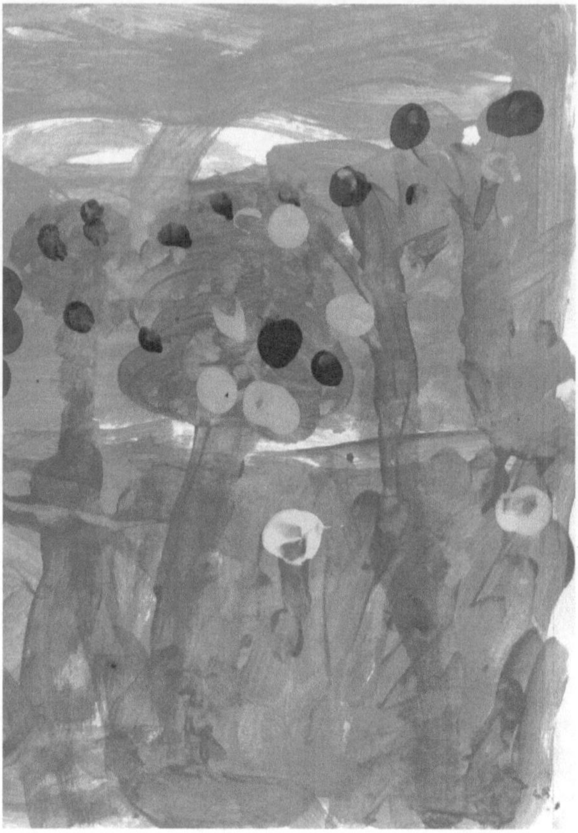

Figure 0.2 An orchard

Source: Sebastian Smith

1 The spaces and power of young people's friendships

Immersive geographies

> I have a dream that one day down in Alabama with its vicious racists, with its governor having his lips dripping with the words of interposition and nullification, one day right down in Alabama little Black boys and Black girls will be able to join hands with little white boys and white girls as sisters and brothers. I have a dream today.
>
> (Rev Dr Martin Luther King, Aug. 28, 1963: speech delivered as part of March on Washington)

> If our children grow up playing and learning with people from different backgrounds, they will be less prejudiced, more understanding of difference, more confident and more resilient, living in a globalised and connected society.
>
> (The Casey Review into Opportunity and Integration, 2016: 54)

This book foregrounds young people's social experiences in education, and I hope that teachers, pedagogues, social workers, parents and others with an interest in forging young people's subjectivities will engage with its findings. The book makes one critical statement: that young people's social worlds are at least as important to their experience of, and engagement with, education as any formal curricular activities, and are potentially transformational in the ways in which identity categories, which frame a host of dis/advantages and inequalities in societies, are understood and performed relationally. The transformative potential of young people's socialities is often evoked by a range of political commentators, as the two aforementioned quotes, from Rev Dr Martin Luther King and Dame Louise Casey, typify. Following this lead, is it a step too far to argue that young people's social relationships can and do sometimes change and transform disability to ability, gender to mutual recognition, race/ethnicity and religion to an irrelevance, except to an appreciation of the richness and resources that religious, cultural and ethnic diversity bring? Perhaps this claim is overly ambitious; these social relationships are reproduced within contexts of broader social relationships and power. Yet friendships forged across diversity by young people provide imminent potential to transform enduring differences, both in the contemporary moment but also in ways that might endure and transform future societies.

In this book, then, I seek to critically deploy the 'futurity' that is embodied in children and young people, who are frequently discursively, materially and politically

DOI: 10.4324/9781003028161-1

deployed as the 'future' of societies whilst so often simultaneously having their embodied experiences sidelined. I foreground the power of young people's socialities, agencies and subjectivities in schools, a power which is often overlooked. In the book I examine how young people's own subjectivities and agencies are powerful in reproducing or transforming enduring differences in school spaces. I argue for the importance of *geographies* of young people, to foreground how young people become and emerge within specific spatial (and social) contexts, and how they forge and remake spaces through their being in the world individually, collectively and in relation to others.

The chapter sets out the context and key ideas of the book. It starts from two critical positions: first, the potential (and frequent failure) of schools to transform societies through the futurity of young people; and second (and relatedly) that young people's social worlds are critical. They are at least as important to their experience of, and engagement with, education as any formal curricular activities, and yet they are often overlooked. In addition, schools do more than 'teach' subjects; they are powerful spaces in which particular (sexed, gendered, dis/abled, racialised, classed) subjectivities are formed. The book takes an intersectional approach, but disability, or bodily-mental-emotional difference and labels of Special Educational Needs are foregrounded, and also race/ethnicity, class and sometimes gender and sexuality, and their myriad intersections. Young people's social relationships are critical and potentially transformational in the ways in which powerful categories, which frame a host of dis/advantages and inequalities in societies, are understood and performed relationally and can, potentially, be challenged and changed. To examine these central contentions, I set out two key overarching original ideas: first, that young people are *contextual bodies/subjectivities/agencies*; and second, the idea of schools as being *immersive geographies*. Rather than focus on young people as individual coherent wholes, I develop the idea of young people as dynamic, porous and connected bodies/subjectivities/agencies. Four central insights emerge from this position. First, the embodied, material, nature of young people's agencies. Second, the contextual and dynamic nature of young people who *become* differently in different social, spatial, historical, political, economic and cultural contexts. Third, young people as nodes of the intergenerational reproduction of enduring differences. Fourth, the powerful nature of young people and specifically their socialities, and their power to reproduce but also to challenge and change enduring broader-scale inequalities via their everyday performances.

Immersive geographies is a pivotal idea in the book. This concept examines the ways in which the coming together of (young) people through repeated encounters in space and time provides a potential for transformation of the ways in which difference is performed, enacted and understood. This difference can be tied to enduring axes of power, such as class, race/ethnicity, religion, gender/sex, sexuality and my specific research interest of dis/ability, along with how these intersect, alongside more subtle differentiations. I reflect upon what is specific about school spaces (although it might be pertinent to other spaces) in which people, and in this case, young people, come together, not just occasionally and fleetingly but repeatedly and enduringly. Time is key here, along with space, and the space/time/space/time/space/time dialectic of repeated encounters in spaces which are ostensibly the same and yet performed slightly differently every time. The bringing together of young people in specific spaces repeatedly allows them to forge deep and affective connections which transform them in some way. The idea of immersive learning is often applied to generating new worlds via virtual reality technologies. This sense of new open possibilities, and the ability to generate new worlds, is critical to immersive geographies. In schools (young) people converge to do similar things day after day, and yet every time they come together the connection, and the practice, is a performance; it is provisional, and new realities are made every day. These immersive geographies are open, porous and connected. They happen in specific places (Massey, 1993, 2005), which are situated within constellations of powers and resources, which constrain (although do not determine) the potentials of the young people's powers. More hopefully, this very connectedness provides opportunities for some of the more empowering ways in which subjectivities are played out to have an impact beyond the immediate space/time of it happening.

Immersive geographies have empowering potentials beyond the immediate space/time of the schools studied. I have two key ideas of how this is so. First, the young people as *contextual bodies/subjectivities/agencies* have trajectories beyond the school as they move into other spaces and times. We can hope that these new radical ways of being together become part of the young people's habitus, a beyond-conscious backdrop to their future social engagements. So, as the young people move through space and time, they are taking these experiences with them in conscious or beyond-conscious ways.

Second, the book is an empowering project. Drawing upon Katz (e.g., 2001, 2004), I argue that the book is a countertopographical project. The book demonstrates how young people in different schools experience similar processes of subjection around powerful and enduring axes of power.

The book presents the real, visceral emotional experiences of young people who are often reported in the news. For instance, news headlines discussing young people with Special Educational Needs or Disabilities (SEND) state that:

> Pupils with SEND accounted for nearly 45% of all permanent exclusions and 43% of fixed-term exclusions in 2017–18, despite accounting for only around 20% of the pupil population.
>
> (Adams, 2019)

Or that: "children with SEND are 5 times more likely to be excluded from school" (Children's Society, 2022). These statistics are shocking; however, statistics belie the embodied and subjective experience of these young people and their families. The accounts of young people in this book humanise that number. In this book, I try to present how it feels to be the young person who is the target of adults' over-surveillance and is kept behind in the classroom because they are seen as 'naughty' when their mind-body-emotional capacities intersect with (racialised/ethnicised, gendered and classed) norms of bodily and mental deportment. I represent the experiences of young people who are left out and isolated in social settings. In addition, I emphasise the resilience and strength of these young people, and how their social collectivities can contest and transform society. Of course, no two young people will experience these processes in exactly the same way; however, this point is of little import; these things are happening to *real* people and behind the headlines are subjective, emotional children and young people.

In general, as adults we feel an urge to protect children, and yet we turn away from and fail to address the experiences of many young people, particularly those with SEND, Black[1] children and young people (if we do not identify as Black), poor children, and other marginalised children and young people, and, of course, these characteristics intersect. Katz (2018) discusses how some young people are cast as 'waste'. Katz does not mention labels of SEND, although this is one way in which young people are labelled, have low expectations attached to them and are set aside for a different, and less successful and fulfilling, life – especially if they are also Black, from a minority ethnic (or global majority) group, poor and/or from a relatively uneducated family (see also Holt et al., 2019a). This book is a deliberate and conscious attempt to foreground the experiences of such young people, to challenge their invisibility and powerlessness, to bring forth their lively subjectivities so that it is difficult to turn away from such astounding statistics.

More critically, the book foregrounds moments in which young people exceed and challenge the exegesis of power, and become something else, radically decentring regimes of dis/ableism, class power, sex/gender and racial and ethnic inequalities. For the most part, these ways of being are moments in space and time unconsciously performed and were just about playing – playing something completely separate and independent from 'real' life or playing with subjectivities and ways of being in the world. The immersive moments with transformative potentials to be otherwise were largely not self-reflected upon by the young people. They have been taken away and analysed by me and others. I have also reflected upon the conditions of the emergence of these moments to examine the possibilities of recreating these conditions in other space/times, providing opportunities for transformation. As such I hope that some of these moments that exceed the exegesis of power that constrain powerful subjectivities, and the conditions for that exceeding, inspire similar ways of being in other contexts beyond those very specific sites in which they were played out. Perhaps more mundanely, there are some good examples in the book of how to encourage more positive connections between young people categorised as 'different' in some way to each other, and it goes beyond just putting them in the same physical space.

This book brings together original, empirical findings from a sustained period of research mostly in schools (also in leisure[2] spaces) in England. The book reflects across five projects with young people aged 7 to 16, which had a broad focus on how dis/ability is constituted and performed through and around heterogenous mind-body-emotional states in schools, and to a lesser extent 'leisure' and home spaces, and how this intersects with other social differences: socio-economic 'class' and gender/sex, and to some extent, religion, race, ethnicity and sometimes sexuality. In total, the research included eight different Local Administrative (Local Authority, LA) areas in England[3]: 182 young people and 100 adults (teachers, parents, educational policy makers and NGO professionals) participated in the research. In-depth observation also occurred in most of the school spaces. In this book, I aim to take a step back and reflect on what new arguments and conclusions can be drawn when all of this work is brought together.

The introductory chapter proceeds through five further sections. First, I indulge in a little autoethnography. Partly this is because my own history and subjectivity are pivotal to my commitment as an educator and a scholar. Partly, I argue that this provides insights into the embodied and contextual nature of subjectivity, albeit refracted from a self-reflective lens that seems unable to shake off a sense of a coherent whole, self. This is *my* history. Second, I reflect on the potential and frequent failure of schools to transform society, through the futurity of young people. The forging of young contextual bodies/subjectivities/agencies is a type of 'anticipatory politics' (Jeffrey and Dyson, 2021). I move on to set out in more detail the key meta-themes of the book: young people as contextual bodies/subjectivities/agencies and in the fourth section immersive geographies. The final section provides a brief map of the rest of the book.

1.1 Beginnings

As a feminist scholar, my own personal experiences, narrative and life history are, of course, present in the discussions: particularly my role as a mother, a scholar and an educator, but also the child that is still embodied in my adult subjectivity. In this section I provide a brief autoethnography and reflect on how I got here (following Kraftl, 2017; Horton and Kraftl, 2012). As Jones and Garde-Hansen (2012: 8) emphasise:

> We are conglomerations of past everyday experiences, including their spatial textures and affective registers. Memory should not be seen (simply) as a burden of the past, rather it is fundamental to "becoming" and a key wellspring of agency, practice/habit.

It is perhaps impossible to avoid presenting these memories as a coherent narrative, but of course my memories are, as Jones and Garde-Hansen (ibid.) and others have emphasised, as much about an act of imagination, as repeated, regular everyday occurrences become melded into key moments that we think we recall: what Virginia Woolf (1972) calls "moments of being" or perhaps rather less poetically can be described as "moments recalled". Yet they are probably not individual moments, but events consolidated and gathered, layer upon layer from all of the mundane and extraordinary years, months, weeks, days, hours and moments of my past. Through the circular repetition of doing similar things in the same places day after day, moments that seem to have taken place consecutively could be years apart, and no doubt I recall the atypical or extraordinary. These moments recalled are always imbued with the emotions and exegeses of the contemporary moment. Keightley et al. (2012) label this the 'mnemonic imagination'. They discuss how

> [t]he remembering subject engages imaginatively with what is retained from the past and, moving across time, continuously rearranges the hotchpotch of experience into relatively coherent narrative structures, the varied elements of what is carried forward being given meaning by becoming emplotted into a discernible sequential pattern.
>
> (p. 43)

On a recent visit back to my childhood home, I found an old diary from when I was 16, and I was shocked by what I had been writing and my concerns and preoccupations of the time. I was deeply concerned about friends who I can now hardly remember and in some cases do not recall at all. I remember being a serious, studious teenager – the 'English, Geography and Biology square' as I was labelled during my A levels. I scooped up the academic prizes, and I certainly was studious and determined. I was also, clearly, a rather frivolous, somewhat vain, but also insecure 16-year-old. The daily occupations of my former self are of little relevance here (and I am not sharing); however, they do cast some doubt on the narrative of myself that I have constructed.

Whilst no one can perhaps truly know how they appear to others, I believe I present as a middle-class white heterosexual non-disabled married woman, a mother of three children and a university professor, of which I am inordinately proud. Perhaps I retain a little of what might seem an estuarine Kent or Essex working-class twang and a certain deferential and self-depreciating manner, which betrays something of my past. Yet I am a third-generation migrant who has had a hybrid life. Most of my great-grandparents were Irish migrants, and this much-diluted heritage was present in the form of compulsory Catholicism, a huge extended family and lively wakes for the recently dead.

My early childhood was in a lone-parent family in a mining village, attending a primary school where having a pro-education ethos and a thirst for knowledge was by no means the norm and sets you up for rough treatment in the playground and being shouted at in the street for being a snob – a label I wore with pride. I also had some lovely friends. Bullying, name-calling and aggression were pervasive, and the name I was regularly called was a racist slur – not because I am Black but because I then had olive skin. Reflecting on this is shocking indeed and says something about the general macho, racist, sexist and homophobic norms which pervaded the 1970s and 1980s in the UK and were stark in my insular mining village community, which was overwhelmingly white and uniformly working-class. Only one teacher lived in our village, in the 'big house', and she was regal in her Rolls Royce, in which her husband drove her to school like a chauffeur whilst she sat in the back. It was a tough, hard place, but there was solidarity and kindness abound. A northern enclave in the south of England. Hostile to outsiders, tough on insiders, but always pulling together for our own.

I was the second person in three years from my primary school to pass the Eleven Plus to go to the nearest girls' grammar school, which was 12 miles by train. The other person was my older sister. The fact that both my sister and I passed the Eleven Plus is testament to my mum, who inspired in us an interest in learning and a curiosity about the world and helped us to practice for the examination. I remember asking my teachers for extra homework, something which bemused both them and my classmates. The fact that my rebellious sister was subsequently politely asked to leave might not have made my experience of grammar school any easier. Stark memories intrude from grammar school, including those of looking at the board in maths in my first year and wondering what one number was doing on top of another; I had never seen a fraction. I have memories of some kindness and support from teachers, and the other girls, but also some outright discrimination. At my first parents' evening, one teacher assured my mum and my soon-to-be step-dad Grahame (or as I prefer to call him, dad) that although I was behind, with my work ethic, she predicted I would be top of the class by year Nine. Another teacher possibly inspired my disciplinary choice of geography by telling lies about me to my mum and Grahame and telling me to give up geography as soon as I could. I am not sure whether it was this awful teacher or the wonderful Mr Fox who most inspired my choice, although more positively Mr Fox inspired a creative and questioning mind. My dad is a quiet man, but he queried the untruths this teacher was uttering, and this moment demonstrates the impact of the cultural capital that my second dad, Grahame, as a chartered accountant, university-educated, music-loving, slightly obsessive and hard-working man, brought to the family. As well as the love, of course.

A dramatic shift happened after my mum married Grahame and we moved to a new area away from the mining village to a middle-class village and a new, wonderfully nurturing girls' grammar school (Barton Court) where I was accepted and encouraged by the other students and the teachers and where I acquired my more middle-class accent. In my new school, my amateurish attempts at teenage rebellion were met with kindness and questioning. My teachers were almost without exception wonderful, supportive people. I could name any number of teachers who were pivotal in supporting me in what was, after all, a turbulent change.

Of course, in my own narrative, my enduring interest in education and in/equality is tied to my own experiences, some of which I have shared here, whilst others are too personal to share. Most importantly, despite our many problems, despite some very early hardship and adversity, my home was full of love, and I have had some wonderful friends. My mum is a force of nature, and despite leaving school at 15, she went back to university and graduated in the same year as me. (After a 20-plus-year career as a social worker in which she has touched and improved countless lives, she is now retired and has written two novels. She is 71 and still plays tennis many times a week.) Before we had a car we travelled around on trains and buses. We rarely spent a free day at home. We went to libraries, to London to the museums, to the coast: everywhere! I had great friends on my street, and we used to have immense freedom to go around the village; although I was strictly forbidden to leave the village, we once cycled around 12 miles to the coast.

When we moved, I made new, wonderful friends at my school and home. With Marian I shared my love of historical architecture, performance and music – although the performance was the talent, for me, not the music. With Helen and 'the gang' we went to gigs and, later, some incredible house parties held by the local university students. Then off to Leicester University for my BA in Geography, which was just the start. I will end it there, because this book is about childhood, and I wanted to reflect on how mine has forged who I am. (Sorry Nicholle and Kate, you are for another time.) Training to be a teacher, living in Portugal and coming back to study for my PhD are not relevant here. I wanted to reflect, for a moment, on the transformative opportunities of education. This was something that my mum was keenly aware of back in the mining village. I wonder how my history of migration, family reformation and changing schools made this dream possible. I am acutely aware of the power that education has to transform lives in multiple ways, and also of how often it does not.

1.2 The potential (and failure) of schools to transform societies through the futurity of young people

Children and young people are not a blank canvas – they have their own personalities and dispositions and are forged in a social and spatial world in specific ways. But, as Rev Dr Martin Luther King and Dame Louise Cassey, and every teacher and every parent probably believe, childhood and youth present a unique opportunity to recreate the world in different and more positive ways. This puts a lot of responsibility on children to overcome the messes of the generations before them. Nonetheless, it is a taken-for-granted position, which, as I argue through the examples in the book, has some validity. What is less broadly understood, perhaps, is the ways in which young people so often reproduce, repeat and re-enact all of the hierarchies, prejudices and power relations which they learn through their education in all its forms, within the societies in which they are embedded. Despite the importance placed on young people's co-presence within schools to overcome entrenched societal problems tied to diversity, there is a common failure to engage with young people's own social worlds within school spaces and the potential therefore of young people's social relationships to be a key factor in addressing (or failing to address) educational inequalities.

Many social commentators, politicians, development organisations and governments state that education is a key mechanism for promoting social mobility and addressing inequalities, deploying the futurity of childhood and youth. Enhancing educational opportunities and reducing inequalities in education is a key strand of most development policies, as enshrined in UN Sustainable Development Goal Four: "to ensure inclusive and quality education for all and promote lifelong learning" (United Nations, 2015). In stark contrast to the fact that politicians of every persuasion point to the transformative power of education to improve lives, promote social mobility and reduce inequalities, educational inequalities persist. These inequalities, between wealthier and more educated groups, with higher levels of cultural[4] capital, and poorer and less educated groups, and along racial and ethnic grounds, are persistent and stubborn.

There appears to be an interesting paradox by which education simultaneously enhances the quality of life and opportunities and (re)produces existing patterns of advantage and disadvantage within societies. For instance, Raudenbush and Eschmann (2015) emphasise that expanding education reduces class-based inequalities and provides general benefits to societies (see also Downey and Condron, 2016). However, there is a substantial body of literature that documents how education reproduces enduring advantage and disadvantage. This is pronounced within education systems which are characterised by neoliberalism, incorporating competition, 'choice', marketisation and fiscal responsibility (e.g., Sirin, 2005; Ball, 2003, 2013, 2018; Calarco, 2014; Ball and Vincent, 2007).

A recent UNICEF Innocenti report card highlights huge disparities in educational attainment at all levels in the richest countries of the world, including the UK (UNICEF, 2019).

Given the stated political prioritisation of education, it is imperative that the mechanisms for these inequalities are understood and destabilised. It is, perhaps, unsurprising to find that fee-paying, independent schools are a mechanism for the reproduction of the privilege of the affluent and educated classes (Ball, 2013). More surprisingly, perhaps, critical scholars have long emphasised that state-provided school education is often pivotal in the (re)production of socio-economic inequalities based on class or socio-economic status, race/ethnicity and the maintenance of middle-class and white privilege (Ball, 2003; see also Jenkins et al., 2008; Calarco, 2014). These patterns have a spatial expression: state school-level education is segregated around class (Harris and Johnston, 2008) and/or race, and ethnicity (Khattab, 2009; Modood, 2004; Burgess and Wilson, 2005) and, indeed, 'ability' (Gibbons and Telhaj, 2007).

Within investigations of why schools often reproduce, rather than challenge and transform enduring inequalities, there is discussion about the relative importance of internal versus external factors. Wilson and Urick (2021) identify the importance of providing opportunities to learn. Anderson (1982) developed the concept of 'school climate' to conceptualise "the psychosocial school atmosphere, and intergroup interactions which affect student learning and school functioning" (Maxwell et al., 2017: 2). By contrast, other scholars point to the importance of the socio-spatial context in which the school is situated and ultimately the resources, students and teachers that are within the school space. For instance, Gibbons and Telhaj (2007) argue that the best-performing high schools become and remain high performing by attracting students from families with high levels of cultural capital who perform well according to the bourgeois norms of the testing regime, leading to a virtuous cycle of success for these high-performing schools (see also Gibbons and Telhaj, 2016). Meanwhile, Ruth Lupton and Hayes (2021) have emphasised that socio-economic disadvantage has a connection with high levels of Special Educational Needs and Disabilities (SEND) and that this exacerbates other contextual impacts of a disadvantaged student population.

The role of young people's own social practices and relationships as a mediator of patterns of educational attainment has been somewhat relegated in these debates, although there are pointers towards the importance of the students themselves and their own characteristics. For instance, Berkowitz et al. (2017) emphasise that the characteristics of the students and the population that a school services have substantial impacts on the "school climate", whist Domina et al. (2017) acknowledge the role of young people themselves in influencing the ways in which schools act as 'sorting machines' which create 'categorical inequality', generating

> meaningful social categories by deciding which students to enrol and by repeatedly sorting students into age grades, ability groups, and instructional tracks, among other formal and informal groups.
>
> (p. 314)

In this book, I examine many axes of difference and how they intersect. Particular attention is paid, however, to dis/ability: disability and the intersecting label/experience of Special Educational Needs and Disability (SEND). Being disabled or having a SEN label/experience is the characteristic that leads to the highest levels of disadvantage in schools. The school-level education institution is, in entrenched ways, ableist (McGillicuddy and Devine, 2020); in this context young people with SEND labels have the worst educational outcomes of any social grouping (see for instance Azpitarte and Holt, 2023). SEND intersects with other identity characteristics, such as class and education level of families, gender, race/ethnicity and so on. Black boys from poor backgrounds, and indeed any group from poor backgrounds, or those with low education levels are extremely disadvantaged, particularly if they are also labelled with SEND, which they disproportionately are (Banks et al., 2012; Dyson and Gallannaugh, 2008; Youdell, 2010; Cruz and Rodl, 2018; Strand, 2016; see also Holt et al., 2019a and Azpitarte and Holt, 2023 for a more extensive discussion).

Some of these categories have wider relevance outside of school spaces (e.g., class, racialised and gendered categories). Others are, perhaps, specific to school spaces, such as age grades and ability groups. Domina et al. (ibid.) emphasise that students internalise representations of themselves as learners and adjust their expectations and performances accordingly (see also Benabou and Tirole, 2003).

Within the context of the global Covid-19 pandemic, these inequalities have been writ large and exacerbated on a variety of intersecting scales, from the global (between regions and nations, with poorer areas more badly affected and disrupted) to the local and even individual – with girls, young people from poorer families in more disadvantaged areas, young disabled people, and those from Black and minority ethnic backgrounds being most disadvantaged in most contexts (Cortés-Morales et al., 2021; Holt and Murray, 2022; Franklin and Brady, 2022; Hunt et al., 2022), but particularly in those nations and areas which were most badly affected by Covid. The nations worst affected by Covid were largely those with high pre-existing inequality and with intensely neoliberal and populist governments, including Brazil, Chile, the US and the UK (McKee, 2020). Despite the global scale of the pandemic, the impacts were vastly different in diverse contexts. Differences were evident both in relation to morbidity and mortality (experienced most often by young people as bereavement) and the level of disruption tied to interventions on mobility, movement, social connection and education.

There has been significant disruption to education, caused by school closures during lockdowns and as a result of local cases of Covid-19. To take two examples: in the UK, in addition to the minimum of six months of education lost to most young people during two national lockdowns (March–September 2020 and December 2020–March 2021) news headlines at the end of June 2021 outlined the crisis of 279,000 young people being absent from school after coming into contact with someone who tested positive for Covid-19 (Richardson, 2021), whilst in Brazil, a study released by UNICEF points out that the country had almost 1.4 million children and adolescents between 6 and 17 years old out of school. It also concludes that more than 5.5 million Brazilians in this age group did not have school activities *at all* in 2020 because of the pandemic (UN News, 2021). These absences, as young people move through their school trajectories, present a serious issue for education from both a learning and social perspective.

The UK makes an interesting and pertinent case study, for all the wrong reasons. It is the fifth wealthiest country in the world based on overall Gross Domestic Product, and yet according to a recent UNICEF report, it has high levels of educational inequality (UNICEF, 2019). As the Education Policy Institute states:

> The most persistently disadvantaged pupils . . . are almost two years (22.6 months) behind all other pupils by the time they finish their GCSEs.
>
> (Andrews et al., 2017: 10)

The context of Austerity measures within the UK, which reflected those in the EU and in other geographical contexts following the 2008 financial crash, was significant, with its associated cuts in public spending and services. In the UK Austerity was a political choice, yet in many indebted countries across the world, Austerity measures are imposed by the IMF. Even in the UK, progress towards decreasing the divides between disadvantaged and other students has been variable; the gap between the most disadvantaged pupils and their cohort has increased by 2.4 months (Andrews et al. ibid.).

1.3 Young people as contextual bodies/subjectivities/agencies: reproducing and challenging axes of power

Within all these debates about education and the social reproduction (or transformation) of inequalities and disadvantages, there is often focus on formal aspects of education, the acquirement of symbolic and cultural capital in the form of accredited qualifications, and 'success' in the form of transitioning to good employment or further and higher education. These aspects of education are, of course, critically important and unequal. Yet schools do much more; they teach young people to be appropriate citizens (Pykett, 2007; Vincent, 2019; Mills, 2021) and to control and regulate their emotions appropriately (Gagen, 2015). These (ab)normalising ideas (Holt et al., 2012) of appropriate deportment, emotional regulation and becoming good citizens is constituted within understandings of what makes a good citizen in any specific context and can draw upon and marshal particular discourses of the nation (Gagen, 2003). These discourses of the appropriate, self-regulated citizen are also tied to and reproduce gendered, classed, racialised, sexualised and religious norms (see for a recent example Hall, 2020; also Maguire, 2021; Martino, 1999; Connolly, 2002), and demonstrate how schools are sites of governance, regulation and normalisation that survey, monitor and trace young people at the most intimate level. Within neoliberal education systems which put a high purchase on competition and also normative standards of education, such as in the UK, schools are becoming increasingly ableist institutions, particularly around learning and emotional 'ability'.

The *role of the agency of young people themselves* to social reproduction has been under-explored and is often taken for granted. Young people's own potential to transform enduring inequalities is too often assumed rather than interrogated in full in policy and practice. In discussions of educational inequalities, young people's socialities are often relegated to 'peer effects', which tend not to be accompanied by detailed examination of children's own social and geographical worlds.

By contrast, almost without exception, geographies and social studies of young people emphasise that children are social agents.[5] Recent commentators have emphasised a need to critically interrogate the nature of children's agency (Holloway et al., 2019), and Kraftl (2013) has pointed out the need to go beyond agency (see also Kraftl, 2020). Nonetheless, the political and conceptual centrality of young people's agency, as powerful actors who act meaningfully, remains unchallenged in geographies and social studies of childhood, and is arguably crucial, given that children and young people's political potential has often been sidelined with troubling consequences. Even if we want to decentre and position young people's agencies within their broader social, spatial and material contexts and connections (Kraftl, 2020), scholars within the field retain the view that young people have agency. As Sarah Holloway, Sarah Mills and myself (Holloway et al., 2019) argued, this centrality of young people's own agency retains a political and conceptual imperative in a global context in which "there are numerous time/spaces where young people are denied agency, rights and/or participation" (p. 472).

Within the subdisciplines of geographies and social studies of children and youth, we can congratulate ourselves and even challenge this paradigm shift (James et al., 1998). Nonetheless, many of the most powerful disciplines that forge and change young people's lives – from social policy and social work to educational and developmental psychology, to much of educational studies and pedagogic development – have at best partially engaged with children and young people's own agencies, and at worst often completely overlook children's own voices, agencies and their rights. In an example of the latter, Bessell (2017) critiques the continuing use in social work of the 'strange situation' approach to assessing a young child's attachment to their carers, in which a carer leaves a child with a stranger for a short period of time. This method, common in social work, highlights that even in a discipline and profession so central to children's experiences, it is considered necessary to temporarily distress a child for the purposes of an assessment, for what is seen as their greater benefit.

In Holloway et al. (2019), we argue that there is a tension between post-structuralist feminist destabilising of agencies and this political imperative to foreground young people's own agencies. On reflection, I don't quite agree with setting up this tension. The kinds of post-structuralist and feminist debates I draw upon do not destabilise the notion of *agency*; rather, they question the notion of a sovereign, fully rational, all knowing *agent:* a masculinist, ethnocentric, ableist concept of a coherent agent which has been used to deny agency to those whose only way of being in the world is inherently and obviously forged in interdependency, including young people, disabled people, certain racial or ethnic groups and often women. These kinds of approaches also emphasise that structures are not outside of, or a predeterminant of, agency, but that agencies produce structures which do not then exist outside of, but subsequently intersect and forge, agencies within these contexts of 'structures' which might seem predeterminate, yet which are always being forged, with the potential to be created differently.

Working productively across the fields of geographies of children and youth, and feminist and post-structuralist perspectives, I position young people as *contextual bodies/subjectivities/agencies*. This emphasises the importance of four key points. The first is the embodied nature of young people's agencies. The second is the contextual and dynamic nature of young people who *become* differently in different social, spatial, historical, political, economic and cultural contexts. The third is understanding young people as nodes of the intergenerational reproduction of enduring differences. Fourth, I emphasise the powerful nature of young people and specifically their socialities and their power to reproduce but also to challenge and change enduring broader-scale inequalities via their everyday performances.

As contextual bodies/subjectivities/agencies, young people are *embodied* – the materialities and the specific dynamism of young people's bodies matter: their nature, their genes, their corporeality and the reality that whilst all bodies-minds are dynamic, young people's bodies are growing, developing and changing at a specific rate. Childhoods, more than any other stage of the life course, are foundational and stay with us as memory, as habitus, as tastes, which are challenging to transform, and as corporeal matter. Yet, young people's embodied subjectivities are *contextual*: material bodies-minds and emotional states are not pre- or indeed post-social; they emerge in a constant dialogue and becoming between material matter or 'stuff' and the socio-spatial contexts of young people's lives – both in the present and in the past (Bourdieu, 1984, 2018; Reay, 2004a). Young people are *nodes of the intergenerational reproduction of enduring differences*. They are subjects, because their potentialities and ways of being in the world are forged within power; axes of power – expected horizons, social expectations and accepted practices and performances – frame their potentials. These are dynamic and can be challenged and changed; nonetheless, they are powerful 'structures' on young people's horizons and are oft reproduced, frequently in ways that are not deliberate or purposeful. As young people play out and perform their own and others' subjectivities, they often are performing and re-enacting axes of power relations, which connect to and reflect enduring inequalities in society around dis/ability, class, race/ethnicity, gender and other social differences. Material contexts and capitals also matter.

Within these contexts, and constrained and forged by them, emerge young people's (or anyone's) agencies. Yet, young people's agencies can exceed and potentially transform these existing relations of power. Young people are interconnected, interdependent, forged in relation to others (people and things). But they really matter. They are powerful. Young people make sense of, reproduce or even change the world around them, including through their own personal and collective educational journeys. When young people come together in space, it is always an improvisation and a remaking of the world; this can lead to new horizons and potentials, new ways of being that are different and challenge enduring axes of difference. Young people's social relationships are powerful and potentially transformative.

In this book, I focus on dis/ability, poverty and social class, and to a lesser extent ethnicity/race, gender, sexuality and some of their intersections. The role of children's own social worlds in reproducing, and their potential to transform, these enduring differences are oft alluded to but seldom thoroughly interrogated, as exemplified in the two powerful quotes that opened the book. There is often a generalised assumption that co-locating children and young people from different religious, racial or ethnic or class backgrounds, or with and without mind-body-emotional differences or impairments, will lead to a more inclusive society in the future.

Such assumptions do not take account of young people's own embodied social practices, hierarchical social geographies and performances of inclusion and exclusion, sameness and otherness, distance and proximity, belonging and lack of belonging. Whereas co-location might provide a context in which young people become less prejudiced and more understanding of difference, the opposite might also occur. Young people can reinforce tribal differences based on a line across a road, with very real consequences in terms of violence or even death (Brotherton, 2015). An important and central question is, in which contexts does proximity lead to 'encounters' (Valentine, 2008; Ahmed, 2013; Staeheli, 2003) which transform or challenge enduring and negative perceptions of 'the other', and in which do micro-geographies of exclusion and negation continue?

In seeking to begin to address this question, the book reflects on a sustained period of research across five projects with young people aged 7 to 16 (see also p. 2). These studies were subtly different, yet had a common thread to understand the ways in which dis/ability is performed and (re)produced in school spaces, and how this intersects with other axes of power: notably, although not exclusively, socio-economic status, class and cultural, economic and social capitals. The majority of the research involved in-depth qualitative research with children and young people, including semi-structured interviews, ethnographic observation and some more creative methods, such as self-directed photography and even a street dance.

These methods sought to understand children and young people's social worlds, given that young people are not merely passive recipients of the expectations of adults around them. Within any social group of young people, there will be subtle plays of power, hierarchies, projections and connections, empathy and distanciation which also connect to micro-geographies of inclusion and exclusion which may or may not be observable to adults around them. For the most part, unless they are particularly problematic, adults do not focus in detail on young people's social worlds, and the importance of young people's friendships and sociality to their experience of school is often underplayed by comparison to formal pedagogic considerations. It is therefore crucial to consider young people's agencies.

1.4 Immersive geographies: potentials to be otherwise, tendency to the same enduring social differences

This book focuses on the coming together of young people in schools and the political potentials this has to transform and challenge enduring social inequalities and disadvantages. Schools are particular and special spaces, with specific characteristics: sites dedicated to learning, which is inherently political (James et al., 1998), and what is deemed as important to learn is tied to the political exegesis of any given time, as demonstrated by the Gradgrind[6] curriculum of the UK Conservative-Liberal Democrat Coalition and the Conservative government since 2010. Schools do more than teach specific curricula, however; they also teach ways of being in the world and subjectivities either in deliberate or less purposeful ways.[7] For the purposes of this book, schools are interesting because they are the site of a coming together of people from different families, sometimes from different backgrounds, to be alongside each other repeatedly over time, and in such a concentration of people of a similar age as to be unique. That the people who are thrust together in this way are young people, with all the futurity and expectation placed upon them, who are also growing, developing, learning, makes schools exceptionally interesting sites of study, with imminent political potentials.

Clearly, the expectation that co-locating young people will transform entrenched and deep-rooted 'axes of power' is a simplistic perspective which fails to fully take account of young people's agencies; not to mention that young people's agencies and subjectivities emerge within specific spatial and social contexts which are connected to broader media, social media and other societal messages that permeate societies in respect of all of these embodied subjectivities. For instance, there are important questions about which young people are coming together in which schools: compare for example ability and class-selective state non-fee-paying schools (Holt and Bowlby, 2019) with disability and class-selective special schools (Holt et al., 2019a) which are differentiated by socio-economic background as well as 'ability' or a label of SEND. Nonetheless, co-location creates potentials and can lead to 'recognition' of differences, and a building of empathy in ways which *can* lead to new connections across difference and performing embodied agencies and subjectivities in new and empowering ways. These potentials for transformation are central to recent geographical interest in the notion of 'encounter'. They are, at heart, about the idea of overcoming social distance through a removal of spatial distance (see Simandan, 2016). Proximity is key. They are also hopeful geographies that anticipate a different and better future (Anderson, 2006, 2017).

Arguably, in geographies of encounter (Valentine, 2008; Wilson, 2017) which are explicitly about specific moments in space and time, space is often prioritised over time (Amin, 2002). Whilst I draw upon the importance of the way things come together in specific spaces and the potentialities of these moments of encounter, I also want to think more carefully about duration. I focus more specifically on the question of *time* within space/time and propose the concept of 'immersive geographies'. Immersive geographies consider the 'depth' of repetition, to be immersed in something, which comes from the repeated and regular meetings of groups of young people (and adults) in schools. They also speak to the immersive technologies that produce new worlds through virtual reality but suggest that young people through their everyday practices are reproducing new worlds every day. Subtly different, perhaps, but each time an improvisation and a different take on script that they have been handed, and which they can challenge and change.

Immersive geographies consider more fully *time* alongside *space*, or as mutually constructive of a performative space, as in Timespace (May and Thrift, 2001).

In figuring immersive geographies, I aim to move away from the potentially fleeting nature of encounters to highlight how children and young people forge social connections by the creation of shared memories through a history of repeatedly sharing social spaces over time. This emphasises a notion of depth, or layers of shared history and repeated connections and an immersive experience of getting to know others. I want to consider more fully the idea of repetition, and repeated and enduring performances, within spaces, which are themselves of course dynamic, performed and worked anew with every coming together of people and things (Gregson and Rose, 1990; Massey, 2005; Doel, 1999). These specific, yet repeated, circular moments of time taking place day after day with the same people in the same place doing similar, yet subtly different, things are not left behind. In immersive geographies, I want to explore how these repetitions create potentials for shared, embodied, emotional and affective collective memories. This could be expressed as space/time/space/time/space/time repeated.

These circular moments of space/time/space/time/space/time also lead to a longer, straighter time trajectory. These shared collective moments become an embodied part of young people's subjectivities, sedimented within the memories and habitual ways of being, of the young people. Young people are forged in these spaces in ways which are, perhaps, never fully left behind and are formative to their subjectivities. In this way, it is possible to take a longer view of time and consider how young people are nodes on the intergenerational reproduction of disadvantages, whilst at the same time, in their specific moments of coming together and the provisionality and improvisation of their performances and play, present a unique collective moment for challenging and transforming these enduring disadvantages in ways that might be taken forward as young people continue on their life trajectories. Through countertopographies (Katz, 2001, 2004) it might be that the contexts of the emergence of ways of being that challenge enduring embodied inequalities via new connections can be drawn out in other schools.

Thinking through the idea of children's shared histories and geographies first struck me in the early 2000s, when I conducted ethnographic research for my doctoral thesis, and I want to foreground one example which crystallised my thinking. In returning to my research diary from my PhD thesis, in which I detail my interpretations of a period observing a class focusing on Alfie,[8] a white British working-class boy with visual impairments and on the Austism Spectrum, I have to reflect that at the time I had relatively little experience with children on the Autism Spectrum in mainstream schools. The fact that I was so observant of Alfie's different behaviour attests to this and stands in stark contrast to Alfie's peers:

> Alfie goes and sits close to where the other children are, but he crosses his legs, and with his hand between his legs he rocks back and forward, not paying any attention to what is going on.
>
> (During this time, the teacher continues to ask questions)
>
> Alfie sits still for about a minute before he is rocking again.
>
> Then he is sitting very close to another boy, who doesn't seem to mind.
>
> The teacher explains to the children what they have to do, and there is a lot of general chat about the work, to the teacher but also to each other as the

children get their things ready. The children seem very excited, happy and interested.

Meanwhile, Alfie is continuing to sit with his head on his hands, then he rests his hands on his chin, then he is clapping again, then he studies his fingers very thoroughly, and then he is counting again.

The other children just don't react to this at all.

Alfie continues as before, as the other children continue getting ready.

All the children laugh about something and after a pause Alfie laughs really loudly.

Then the children move to the tables, and the teacher says they will go to the tables depending on how smartly they are sitting.

[Later] . . . the teacher dismisses the children, saying things such as: 'if you're wearing purple, you can go'. Alfie is in the second half of children to go and line up, with 'people with names beginning with A'.[9]

(Research diary, year four class, Church Street)

1.5 The book

In the book, importantly, I emphasise that young people's social relationships are powerful elements of their experience of school, and that for many, a feeling of belonging to young people's social worlds is pivotal and intrinsically connected to engagement with formal aspects of school and consequently an important component of educational success. This relationship is complex, as unfortunately those young people who are excluded and identified as not belonging within formal elements of school are all too often the same as those who are excluded and coded as 'not belonging' within formal aspects. I suggest that young people's own social relationships are, however, an important and too-often overlooked factor. Encouraging and supporting young people's friendships should be a more central element of school life, especially for those young people who have been, or who are more, vulnerable to being, bullied or marginalised.

At times, the stories and narratives presented within this book are difficult and emotionally wrenching. These are real young people, telling their own stories, of course, mediated and analysed by myself and my colleagues. It is, however, critical to understand and empathise with some of the difficult and challenging stories to build a collective will to change and transform the structural elements of schools which render some young people so isolated and othered. Within the book there are also some light and happy moments, which provide insight into how some young people who had experienced a lack of belonging, or who are vulnerable to exclusion and othering become included, have good friends and a positive relationship with school. These moments provide insights into potentials to design inclusive schools, from every perspective – class, gender/sex, race, ethnicity, religion and dis/ability, alongside their many intersections. Across the stories presented in the book, there are examples of transformative potentials in relation to dis/ability, gender and ethnicity, not, however, so often in relation to socio-economic difference and poverty.

This book has been in gestation for 22 years and was written over a number of years, with a significant proportion written in 2020/21, when the world faced the challenges of a global pandemic. The pandemic brought into sharp relief priorities around young people, and ideas about what school is *for*. There has been a heightened emphasis on mental health and well-being and the importance of social relationships in many media debates and among educational professionals. Nonetheless, after the first lockdown in which children were absent from schools for up to six months, the English government's priority was formal elements of education. I began to write this introduction towards what we hoped would be the end of the third UK national lockdown: schools had been closed since December, and they reopened on 8th March. In the second lockdown schools remained open, despite increasing evidence that older young people transmitted Covid-19. In the third lockdown, when schools had to close again, the English government, concerned that children were falling behind, decreed that every child and young person should be provided with a full timetable of home schooling. I recall the pressure this put on colleagues who were working at night on their paid work, after spending a full day home schooling. This focus on formal education contrasted with the views of teachers, the Children's Commissioner and children's charities who have emphasised that the real cost of young people not going to school is their mental and emotional health, and highlighted the importance of young people's social relationships.

The crucial nature of young people's own sociality and social relationships has been thrown into sharp focus by lockdowns, when young people had limited social contact, and these patterns were reflected in many contexts globally; the pandemic certainly brought into question attitudes and approaches to schooling, education and young people's social relationships. I argue, however, that social relationships need to be viewed as pivotal to school – they are rightly important to young people's mental health, but they are also the very stuff and foundation of young people's schooling, forging subjectivities and educational identities. Young people's social relationships often greatly influence their very connection with and aptitude for school, alongside being a mechanism for the reproduction and potential transformation of broader social and spatial advantages and disadvantages.

This book is structured into nine chapters. Chapter 2 outlines methodologies and approaches which underpin the analyses in the book. By far the greatest priority is given to young people's own accounts of their experiences in schools, alongside ethnographic research within school and non-school spaces, although the stories of young people are not examined within a vacuum and the factors which constrain and enable them and forge their very subjectivities in specific ways are also explored. Some specific ethical problems tied to a conflict between confidentiality and child protection are discussed. Chapter 2 highlights attempts to develop empowering research and details limitations to the approach. The research is underpinned by an empowering approach. By focusing upon empowerment, I want to emphasise the potential of a range of different approaches. This aim focuses on prioritising the most marginalised groups (here disabled young people, particularly those with socio-emotional differences and young people from poor backgrounds), reflecting on ways to enhance and improve their life chances and experiences. Chapter 2 emphasises that the book intends to be an empowering project drawing upon countertopographies to demonstrate other ways of doing difference.

In Chapter 3, I set out why young people's social relationships matter: how and why they are powerful. In addition, it clarifies how young people's agencies are theorised in the book, through the original concept of young people's ***contextual bodies/subjectivities/agencies***. Rather than focus on young people as individual coherent wholes, young people are identified as dynamic, porous and connected bodies/subjectivities/agencies. The importance of young people's own social relationships is situated within the concepts of embodied social and emotional capital. Four key interconnected interventions emerge which develop the central idea of young people as embodied and becoming ***contextual bodies/subjectivities/agencies***, drawing upon insights from Bourdieu (habitus), Judith Butler (performativity, subjection and recognition) and Foucault (subjection, normalisation). The first idea is the embodied nature of young people's agencies. The second is the contextual and dynamic nature of young people who *become* differently in different social, spatial, historical, political, economic and cultural contexts. The third is understanding young people as nodes of the intergenerational reproduction of enduring differences. Fourth, the powerful nature of young people, and specifically their socialities, is outlined, and their power to reproduce but also to challenge and change enduring broader-scale inequalities via their everyday performances.

These themes about the potentially transformative power of young people's social relationships are taken forward in Chapter 4. How and when young people challenge and transform enduring, intersecting axes of power – gender, class, race/ethnicity and, specifically, disability – are examined in this chapter. Here I set out the idea of 'immersive geographies', which captures both immersion and a sense of depth and immersive geographies as open and connected to radical new connections and ways of being. Schools are sites of immersive learning; they are institutional spaces dedicated to educating young people in formal and informal ways, which intersect. Schools are specific spaces in which young people, come together, not just occasionally and fleetingly, but repeatedly and enduringly. These repeated meetings facilitate shared collective histories. Every time they come together the connection and the practice is a performance; it is provisional, and new realities can be created, every day. New connections are forged beyond frames of reference of (dis)ability, class, gender/sex, sexuality, race and ethnicity, religion, via empathy and recognition. These transformations can have implications beyond the immediate space/time of the school, via young people's embodied subjectivities as they move through future space/times and via countertopographies. Immersive geographies are countertopographical projects. These projects seek out alternative ways of being and connecting and by tracing the origins of the emergence of these new connections, suggest ways that more empowering social relationships can be fostered.

The subsequent three chapters illuminate my own (and others') interpretations of the empirical research with young people, and they unpick the embodied, emotional and affective nature of young people's socialities – which are, however, always fraught with power. I explore the potentials for young people's powerful socialities to transform and challenge enduring power relations. In Chapter 5, I examine how young people's friendships provide emotional support, social and emotional capital and 'recognition'. The chapter also witnesses how foundational young people's social relationships are to them, but also how fraught and provisional, the constant work required to make and forge friendships and the dynamism, instability and fragility of friendships, some of which endure despite this need for constant affirmation and work. The ontological need for recognition, which most (young) people experience, means that young people as *contextual bodies/subjectivities/agencies* are always, already intersubjective and interdependent. Friendships of young people, then, provide emotional and social 'capital' and also position them within formative and generative frameworks of power that underpin subjection. Moving forward with this argument, in Chapter 6, I go on to reflect upon the ways that young people's sociality is infused with subtle and more obvious performances of power, demonstrating that young people are nodes of the intergenerational reproduction of enduring difference. The chapter examines hierarchies of friendships, moving on to geographies of exclusion, marginalisation and stigmatisation. Although these geographies represent young people's own rationalities, they often reproduce intersecting, enduring axes of difference: gender, sexuality, race/ethnicity, class and dis/ability. Chapter 7, through the concept of immersive geographies, reflects upon the power of young people to challenge and change enduring realities. Immersive geographies provide scope to enact the power of young people as *contextual bodies/subjectivities/agencies* to challenge and change enduring inequalities. The chapter argues that immersive geographies, where young people come together repeatedly and yet where each time is an improvisation, provide potentials for new lines of flight and new ways of being. These moments are particular moments in space and time, and yet live on in the embodied subjectivities of the young people as they continue on their trajectories. Further, these moments can skip space and time to show different and more emancipatory ways of being. One potential mechanism for this skipping of space and time is, perhaps, this book.

In Chapter 8, I focus my gaze away from the young people to reflect upon some of the ways in which their lives were positioned in relation to broader social, spatial and political processes. It was critical to me to step back from being overly celebratory about the potential power of young people's socialities and to reflect upon schools and young people as porous, connected spaces, variously positioned in broader socio-economic and political processes. The potentialities of young people's subjectivities are constrained and positioned within these broader processes in some stark ways. I was compelled to reflect upon this given my perception that the UK and other countries are currently experiencing a revanchist capitalism beyond anything I could have imagined when I first embarked upon this project. Yet, the young people's accounts also provide countertopographies (Katz, 2001, 2002, 2008). If everything that is global is simultaneously local, it is possible that something that challenges and changes representations and performances of difference in a specific, small-scale point in space and time, can jump scales and have resonance far beyond the specific context of its emergence. Is it possible that some of the more affirmative performances presented in this book provide alternative ways of being and doing that deconstruct difference as otherness or, even more ambitiously, reframe difference and connection around a continuity of mind-body-emotional types?

Chapter 9 draws together the major contributions of the book: the importance of young people's own socialities to their experiences of school, and how young people are critical in the reproduction and potential transformation of enduring axes of difference and entrenched disadvantages around class, dis/ability, race/ethnicity, gender and other differences, which intersect. The way in which the original and new concepts *contextual bodies/subjectivities/agencies* and *immersive geographies* instigate new directions for social sciences and education is considered. As *contextual bodies/subjectivities/agencies*, young people are embodied yet contextual and dynamic in nature: young people *become* differently in different social, spatial, historical, political, economic and cultural contexts in interaction with their corporeality. As *nodes of the intergenerational reproduction of enduring differences*, young people frequently reproduce axes of power relations, which precede them and continue through time and space, and through which entrenched intergenerational differences continue. Yet finally, the book emphasises the *powerful nature of young people* and specifically their socialities and their power to reproduce but also *to challenge and change enduring broader-scale inequalities via their everyday performances*. The concepts of embodied emotional and social capital express this powerful and contextual nature of young people's subjectivities and emphasises how they are forged within (and can challenge and change) broader socio-spatial contexts. The concept of *immersive geographies* provides original perspectives about how repeated proximity through space and time provide opportunities for young people to challenge and change enduring axes of power, which has implications much beyond the scope of this book. The book is a countertopographical or empowering project, and by reflecting upon the conditions of the emergence of socio-spatial relations which defy enduring embodied inequalities, some ways that schools can be empowering to young people are suggested. The book concludes with my original poem: A Circle.

Notes

1 Black is capitalised in line with conventions that politicise Black collective history, experience, community and identity.
2 Loosely defined as spaces where young people spend their leisure time, although most of the research took place within formalised activities, such as youth clubs.
3 Local authorities vary in size. They are the local area organisation of education and other local services in the UK and administer around 200–300 schools. Their educational powers and responsibilities have been reduced in recent decades with a move towards centralisation and devolved management of education to educational 'trusts' – an association of schools. Local authorities retain power and responsibility for administering and managing processes pertaining to Special Educational Needs.
4 Which intersect with and are a mechanism for reproducing other forms of capital, notably, economic and social.
5 The field is broad; see for instance the journals *Children's Geographies, Childhood, Children and Society*; the 12-volume Springer book series *Geographies of Children and Young People*, edited by Tracey Skelton; and, for example: Lopes, 2014; Jeffrey and Dyson, 2008; Jeffrey, 2012, 2012; Hopkins, 2013; Holloway and Valentine, 2000; James et al., 1998. The Oxford bibliography, by Kraftl et al. (2022), is also a useful starting point.

6 Gradgrind is an educator in Dickens' novel *Hard Times* who is an archetype of the instrumental educator, espousing the usefulness of facts: "Now, what I want is, Facts. Teach these boys and girls nothing but Facts. Facts alone are wanted in life. Plant nothing else, and root out everything else" (Dickens, 1854: 3). Garner (2013) makes the analogy as teachers warn that truancy will rise in response to the changes that Michael Gove designed for the curriculum when he was secretary of state for education.

7 In relation to national identities, see Benwell, 2014; Millei and Imre, 2021; Moser, 2016; Åkerblom and Harju, 2021; in relation to broader political subjectivities, see Mitchell, 2003; Pykett and Disney, 2016; Vincent, 2019; Mills, 2021; in relation to emotional and social performance, see Gagen, 2015; Bowlby et al., 2014.

8 All participant and school names are pseudonyms.

9 Reporting of what was said in the research diaries is presented with single inverted commas to emphasise that this is unlikely to be a direct quote, but is interpreted by the author of the research diary.

2 Methods, approaches, contexts

In this chapter, I outline the methods, methodologies and approaches which underpin the analyses in the book. The book brings together research conducted between 2000 and 2022, which draws upon a variety of sources, including analyses of secondary sources: policy documents, local and national government websites, data from the National Pupil Database, newspaper articles; and primary research with significant adults (teachers, support workers, NGO members, local and national government officials), together with ethnographic research in schools. By far the greatest priority is given to young people's own accounts of their experiences in schools, alongside ethnographic research within school and non-school spaces, although the stories of young people are not examined within a vacuum and the factors which constrain and enable them and forge their very subjectivities in specific ways are also explored.

The research is underpinned by an empowering approach. By focusing upon empowerment, I want to emphasise the potential of a range of different approaches. This aim focuses on prioritising the most marginalised groups (here disabled young people, particularly those with socio-emotional differences[1] and young people from poor, Black and minority ethnic backgrounds, and their intersections), reflecting on ways to enhance and improve their life chances and experiences. Empowering research approaches are not confined to qualitative, participatory research; quantitative analyses of large-scale datasets can have empowering effects. In this book, the research is, however, qualitative and primary research mostly with young people, and foregrounds their own presentations of their experiences, albeit mediated by my own and colleagues' (and the reader's) interpretations.

In the rest of the chapter, I begin by reflecting upon my broad methodological approach in which I have sought to be empowering, albeit that power is complex in nature and cannot be straightforwardly transferred or given to another. I move on to consider some key points and questions about ethics and the safeguarding of young people, which is of relevance to this book but has wider resonance to research with young people, and perhaps research more generally. Subsequently, I consider the importance of the context of the case-study schools and the particular political moment in which much of the research took place. Despite shedding light on young people's experiences during this very specific time period of New Labour and early Austerity, the research also has resonance with contemporary processes

DOI: 10.4324/9781003028161-2

and practices and how things could be otherwise. Finally, I reflect on some limitations of the research and how I have endeavoured to overcome them, particularly limitations in voices, accounts and empowerments.

2.1 Methodological approaches

My research has endeavoured to prioritise the experiences of young people. From the outset I have been at pains to emphasise that the categories of child or young person are differentiated and there is not an amorphous category of child. Whilst this has been broadly agreed upon, it is still often the case that within research the differentiations of young people are not fully considered. So it might be that social scientists explore questions of race, disability or class, for instance, as a topic. However, in going about our usual research (for instance, about young people's experiences of urban space, journeys to school or whatever the topic may be), we rarely consider all the intersected axes of power which connect and differentiate children and young people, and ourselves. In this book, I have sought to reflect upon some of these young people's intersectionalities, and also to prioritise the experiences of young people who are marginalised – particularly young disabled people, those with SEND and those from poor and racial/ethnic minority backgrounds. As I emphasise in Chapter 3, endeavouring to listen to the voices and experiences of young people is complex, since, in common with all subject/agents, their experiences are mediated and their knowledges are subjective, emotional and partial. Their perspectives are, however, powerful and important.

The research in this book has sought to be empowering. The concept of 'inclusive' research (Holt et al., 2019b) does not, on reflection, go far enough. Whilst the idea of empowerment is problematic, as it seems to suggest a simple transfer of power, in the absence of a more satisfactory label, I will continue to deploy this term. The concept of empowering research moves beyond inclusion and towards a need to transform and challenge the status quo, and it goes back to scholarship by feminist and disability scholars and activists, which first inspired my own. This sets out an important distinction compared to endeavours to be participatory or to prioritise co-production, which is currently a considerable orthodoxy in many of the fields of geographies and social studies of children and youth. A problem with co-production is that there might well be a tendency to listen to the most articulate voices, even when deliberately trying to engage with people who are often marginalised through socio-spatial processes. An attempt to be empowering depends on engaging with the experiences of those who are most silenced and who might often not be forthcoming in our research. In my work, this endeavour has led to engaging young people whose schools might not wish their voices to be prioritised, who might not be the first to put themselves forward and who might behave in ways which can seem troubling and troublesome (Blazek, 2021) both within the context of schools and to some extent to me as a researcher.

This book endeavours to be a piece of empowering work, by demonstrating how young people's everyday practices reproduce the larger-scale inequalities which can be evidenced through statistical analyses. Most importantly, the imminent political potential to be otherwise, to challenge and change these enduring inequalities, is revealed. Through countertopographies (Katz, 2001, 2002), these small-scale challenges have the potential to jump-scale and provide new, alternative futures.

2.2 Methods

The research presented in this book has involved secondary analyses of policy documents, newspaper articles, government websites, school policies, the websites of various charities, the Children's Commissioner and so on. Rather than being a bounded activity, this secondary analysis is ongoing and continuous. My antennae are always on the alert for reporting in relation to young people in schools, and specifically those from disadvantaged backgrounds and/or those with labels of SEND. Additionally, I conduct regular searches of the internet and newspapers.

More directly, the empirical material is drawn from five projects with young people aged 7 to 16. In total, the research included eight different Local Administrative (Local Authority) areas in England. In total 182 young people and 100 adults (teachers, parents, educational policy makers and NGO professionals) participated in the research. Some details of the characteristics of the participants who are cited in the book are given in Appendix One. In-depth observation also occurred in schools, on school trips and in leisure spaces. The data includes approximately 18 months of in-depth ethnographic observation, with four months in Rose Hill and Church Street and six weeks to three months in the other schools, except Seadale High School and the Urban High School, where observation was not agreed by the school leaders. I conducted the observations and wrote the research diaries for Rose Hill and Church Street, and Jennifer Lea conducted the observations and wrote the research diaries for the other schools. The schools were selected for their diversity. Appendix Two gives some details of the characteristics of the schools.

The book provides a reflection across these studies, which, despite their differences of focus, broadly examined how dis/ability is performed and (re)produced in school spaces, and how this intersects with other axes of power, particularly socio-economic status, class, and cultural, economic and social capitals, gender and race/ethnicity. The majority of the research was qualitative and involved in-depth research, including semi-structured interviews, ethnographic observation and some more creative methods, such as self-directed photography, making photo-diaries, drawing, making models and even a street dance. In some contexts, particularly in the primary schools, this involved the young participants selecting and designing their own research methods, facilitated by the researcher. In other contexts, the research was more directed. The former gave insights into aspects of life that we otherwise would not have examined, although the latter also had empowering possibilities. These methods are not discussed in depth here and are part of the accepted cannon of research methods for social and geographical studies with children and young people (see Evans et al., 2017; Holt and Evans, 2016). These approaches, however, could be more often used beyond geographies and social studies of young people to enable educators and others to engage with young people's voices and experiences. There is no reason why these methods and approaches could not be used with other research partners, including adults, although they are much more seldom used in these settings.

In this book, I prioritise young people's social experiences in school. The data for this primarily comes from five studies that spanned 2000–2015, funded by Loughborough University, the University of Brighton, the Environmental and Physical Sciences Research Council (EPSRC) and the Royal Geographical Society (RGS), with the Institute of British Geographers (IBG) Grant and the Economic and Social Research Council (ESRC). The research had slightly different focuses – for instance, the initial PhD research (Loughborough) focused on primary schools and on school spaces within one conurbation, whilst the University of Brighton and EPSRC, RGS and IBG grants focused on a particular urban setting (Hastings) and examined urban regeneration alongside young people's experiences, and particularly those with socio-emotional differences. The ESRC-funded work spanned three Local Authorities in the Southeast of England and explored families' experiences alongside connections between home-school and 'leisure' spaces. Importantly the ESRC-funded project had a co-investigator, Sophie Bowlby, and a research associate Jennifer Lea, who were involved in analysing and conducting the research. Despite slightly different foci of the various projects, the emphasis was always on prioritising the experiences of young people, and bringing this together in this book is a powerful way of exploring these.

Most of the research with young people was conducted between 2000 and 2014, just capturing the transition to the Austerity policies that typified the Conservative and Conservative-Liberal Democrat government from 2010, until it was replaced by the wasteful and irregulated spending of the Conservative government from 2018 onwards (Hillier, 2022). Most of the young people in this book are now adults – it has been a long time in the writing. Their stories and experiences continue to be evocative and important, although the context has changed: it has become considerably worse in the UK for most young people who are not the privileged minority. The stories the young people tell are illuminating and remain highly pertinent to all young people's experiences of exclusion, marginalisation and denigration, but also of hope, of recognition and of transformation.

In conducting qualitative research, we are often faced with the challenge of representativeness. There is also the question of how far can these limited accounts reach? What power do they have? Although relatively extensive for qualitative work, with accounts from young people from diverse backgrounds and contexts, and with complex intersectional subjectivities, this research clearly is not statistically representative: statistical representativeness is not the intention of this qualitative work. Of course, the experiences are situated and contextual, and these young people do not and cannot speak for all young people.

The young people's experiences presented in this book are not statistically transferable to other contexts. However, they are powerful and the power lies much beyond their specific space/times. When we hear in the news that: "children with SEND are 5 times more likely to be excluded from school" (Children's Society, 2022), this statistic is shocking; however, statistics belie the real visceral human experience of these young people and their families. The accounts of young people in this book humanise that number. They show a glimpse of the real human feelings of these young people. In this book, I try to present how it feels to be the young person who is the target of adults' over-surveillance and is kept behind in the classroom because they are seen as 'naughty' when their mind-body-emotional capacities intersect with (racialised/ethnicised, gendered and classed) norms of bodily and mental deportment. I represent the experiences of young people who are left out and isolated in social settings. In addition, I emphasise the resilience and strength of these young people and how their social collectivities can contest and transform society. Of course, no two young people will experience these processes in exactly the same way; however, this point is of little import; these things are happening to *real* people and behind the headlines are subjective, emotional children and young people.

In general, as adults we feel an urge to protect our children, and yet we turn away from and fail to address the experiences of many young people, particularly those with SEND, Black children (if we do not identify as Black), poor children and other marginalised children, and of course these characteristics intersect. Katz (2018) discusses how some young people are cast as 'waste'. Katz does not mention labels of SEND, although this is one way in which young people are labelled, have low expectations attached to them and are set aside for a different, and less successful and fulfilling, life – especially if they are also Black, from a minority ethnic (or global majority) group, poor and/or from a relatively uneducated family (see also Holt et al., 2019). This book is a deliberate and conscious attempt to foreground the experiences of such young people, to challenge their invisibility and powerlessness, to bring forth their lively subjectivities so that it is difficult to turn away from such astounding statistics. The book is also an attempt to examine how to challenge and transform enduring power hierarchies between young people, which radically influence their experiences and potential futures; I seek to trace the potentials to be otherwise in geographies of immersion and to reflect on how these can be reproduced in other spaces and times.

2.3 Ethical considerations: informed consent, confidentiality and safeguarding

All of the research was approved through the ethics and risk assessment procedures of the institutions in which I was working at the time. It might be interesting to reflect upon how those have changed, although it is outside the scope of this book. In all contexts, the complexities of following usual ethical codes with young people were reflected upon, and considerable work was undertaken to ensure that the young people could consent to the research, understand the nature and purpose of the research, and had their confidentiality and anonymity protected.

We explained the research to all of the children individually and went to great lengths to try to ensure that they understood the nature of the research (explaining that it is similar to what they often do in school and that the book would be published and read by an unknown number of people, and so on). Some young people had parents who worked in universities. Others had no concept of what a university is, so we explained that it is a bit like a school, but that sometimes the work we do gets read by other people. We asked questions to judge the extent to which the young people had understood what we were doing and what would happen to their accounts.

The safeguarding role of adults working with young people became more formalised throughout the years of the research, and most Universities in the UK and similar contexts now have safeguarding policies and guidance that need to be followed and negotiated alongside ethical considerations about confidentiality and anonymity. However, even in the first project, I had, after reading Alderson and Morrow (2011) and other texts, reflected upon the potentially conflicting imperatives to preserve confidentiality and anonymity and to protect children. This particularly came to the fore in relation to child protection processes and if a young person discloses abuse or neglect.

There was little direct advice in the social science literature on how to address a disclosure when I started doing research with young people, except to think about what approach you would take and to be prepared (Beresford, 1997). Even now advice as to exactly when protection trumps confidentiality is scarce (Hiriscau et al., 2014). Beresford (ibid.) maintains that the fear of this should not prevent research with vulnerable children. She suggests that the best ethical approach is to encourage children to talk to another adult, and if the child refuses, to explain the seriousness of what they have told you and that it is necessary for you to inform someone else. At all times it is necessary to encourage the child to self-disclose to an adult and, if they will not, to forewarn them that you must do so. Confidences must never be broken without a prior warning and, if possible, the consent of the child concerned. This is in line with psychological and social work guidelines. However, there is evident conflict with confidentiality and anonymity, suggesting that adults' responsibility to protect children remains paramount (James et al., 1998).

In the context of my research, a clear disclosure of abuse never did occur. However, one interviewee indicated that her father had been violent in the home (although she did not suggest that he had been violent towards her). As she (and her teacher) had informed me that she has no contact with her father, and her teacher had discussed the family's history with me, I didn't feel that I needed to act, as the issues were already known. The issue of disclosure seems all too clear cut when you read about it in the pages of books that discuss conducting research with young people. However, it is often more complex, as a sense of responsibility to protect young people conflicts with the rights of young people to confidentiality and anonymity, and yet not in a straightforward way: for instance, when young people experience exclusion and marginalisation in a variety of settings, and we might reflect on whether it is pertinent to intervene. There are many examples in the research, but the two I discuss in the following are prominent in my thoughts and memories.

In the first instance, a boy discussed how he was bullied, and also how the teachers always blamed him for any social incidents with other children. Should I have intervened, and pointed out to the teachers that they clearly expected this boy to have complex social relationships due to his label, and therefore were unaware of the bullying he was subjected to? In another example, in a project I managed (with Jennifer Lea as the researcher and Sophie Bowlby as the co-investigator), it was only years later when I put together the layers of a boy's (Andy's[2]) experience that I realised how he was subjected to multiple exclusions and negative experiences. Once we had transcribed and analysed all of the research with young people, teachers, parents and observations in the schools, and I put these together, I could see how this boy was positioned as not belonging, with visceral emotional impacts in every context, and even, it seemed, in his home life. He was adopted (although I know nothing of the reasons why); it is possible and even likely he had early childhood trauma. His adoptive mum was recently widowed, and was poor and lacked resources, had been struggling with mental ill-health and compared her experiences of parenting Andy negatively to those of parenting a 'normal' child – perhaps usurpingly in a pervasively disablist society. His behaviours challenged the expansive norms of his special school, as he would swear and be 'horrible' to the other young people. He claimed teachers did not like him. His mother had challenged how he was treated by his teacher, and her approaches to the school were not taken seriously and she accepted that this was just something Andy would have to tolerate. In the conversations with both the mother and Andy, we adopted a pseudo-therapeutic role (Parr, 1998; Thomas, 2010; see also Bondi, 2005, 2014a). I also wonder, however, whether we should have done more. As a research team, should we have intervened? Should we have offered to be the advocate of Andy and his mum, and use our education and position to support her in challenging the school? It is also important to acknowledge, however, that Andy's way of being in the world was difficult to deal with for the other children, his mum, his teacher and even for us as researchers, although we were empathetic to his lack of belonging and his exclusion.

Behind the stories presented in this book are many such questions. Although we might try to analyse our research in real time, we don't necessarily get a view from multiple perspectives until we have closely analysed all of our materials. There are important questions about when and how to intervene, which I have seen and have myself executed well and badly, and which are never straightforward. When children and youth's rights to confidentiality and privacy are so often challenged (Christensen and Prout, 2002), it is appropriate to be cautious; however, confidentiality does conflict with the imperative to protect children in subtle ways and there are many ethical questions to be asked.

2.4 Analysis

There were a variety of forms of data produced, visual, ethnographic, artefacts and videos of the participatory activities (videoed with consent). Despite this variety, overall, the key analysis was based on lively materialities turned to text to analyse back within the frame of the representational, which is perhaps a limitation of the studies, but also a pragmatic reality of much research. The field notes were written up into diaries, and in total there are over 1,000 pages of thick, descriptive research notes. Analysis is part of the observation and also the process of writing down and reflecting on the observation. Writing down and reflecting, and jotting notes about interpretations, alongside conversations and discussions with the research team or supervisors, is a critical part of the analyses. Similarly, all the focus groups and interviews with young people and adults were recorded and transcribed. We reflected upon our positionality in the field, but also within the analyses, questioning how our positionality influenced our interpretation of the data.

The visual or participatory methods were primarily analysed to prompt discussion with the young people, in line with a photo-voice method (Delgado, 2015). This practice gave young people the opportunity to comment upon and analyse their own artefacts, and some control over how they were interpreted (Smørholm and Simonsen, 2017). As such the analysis was, at least partly, participatory, and in line with endeavours to ensure that young people have some power over all aspects of the research process (Nind, 2011). At the same time, this also framed and constrained the analyses back to the realm of written representations. Thematic visual analysis was undertaken of photos and other artefacts. With the exception of the plasticine model (Figure 5.1) by Mahal and Jasim, the photos and pictures have been reproduced by my children, to retain the spirit of the themes of the original artefacts but to produce higher-quality images suitable for publication.

The text produced in research diaries and interviews were analysed thematically using a variety of tools, ranging from NVivo to coding up into different data tables using Word, and this was an iterative process of data categorisation and organisation, and examining patterns and differences, which differed slightly according to the context of analyses and the preference of the member of the team doing the analysis. From precursory reads-through and conversations, potential categories for analysis were established. These categories for analysis emerged through a combination of a-prior theoretical codes and more intuitive responses to empirical findings, emphasising the creative and systematic elements of analysing qualitative data.

The analysis was conducted initially on a school-by-school basis, and by the different methods or groups researched. The analysis was primarily thematic, although narrative accounts were also generated by comparing the accounts of individual young people from a variety of data sources. This book represents a meta-analysis, when, after a number of years of analysing discrete parts of the whole of this research, I have had the opportunity to pull together themes from across the projects, and examine both narrative accounts from individual young people's stories and a more abstract view from across the entire research data to establish the key themes, commonalities and differences, when all of this data is put together, spliced in different ways, rearranged again. Given the scope of the data and the nature of the book, it is not possible to list all the major and minor themes that emerged. When put together, however, the data emphasises some major themes which are examined in this book:

- the importance of young people's social relationships;
- how young people often reproduce existing axes of power relations;
- the potentials of young people to forge shared histories and trajectories through encountering each other repeatedly in space and time;

- the imminent and sometimes realised potential towards transformation of 'difference' in young people's socio-spatial practices;
- the importance of socio-spatial contexts and adults' practices in constraining and enabling young people's socialities to transform enduring differences and do things differently;
- how young people's micro-experiences connect to broader-scale, enduring inequalities and advantage and disadvantages;
- intergenerational reproduction of (dis)advantage;
- the importance of how schools and young people are positioned in space and in relation to a host of resources, powers and capitals.

2.5 The importance of context and space-time

Most of the research presented in this book was conducted during the period of the New Labour government and its immediate aftermath in the 2000s and early 2010s. The years of the Labour government are looked upon differently amongst socialist, social-democratic or left-leaning commentators, of which I broadly consider myself to be one. The Labour Party is the UK's left-leaning centrist party. Since the mid-1900s it has spent a majority of time as the official opposition party (i.e., gaining the second-largest share of the seats in parliament), although it has had some notable periods in government, including from 1997 to 2010. The Labour government of the late 1990s and noughties could be closely aligned to the Democrats in the US, and its policies were socio-democratic rather than socialist. The Labour government were implicated in some tragic errors of international policy, in which the US was also firmly implicated, notably Iraq and Afghanistan (a country which was finally failed by Trump, Biden and an obsequious UK administration). These errors are not unique but typify recent UK-US politics in the Middle East (e.g., Syria, Yemen, Libya and so many other places). These horrors and failures of international politics are tied to the geopolitical proxy wars of the US, Russia and others, of which the UK is a minor bit part player aligned to the US. These powers have largely offshored their devastating conflicts to parts of the world from where the populations of these powerful nations remain relatively untouched by the impacts (Hughes, 2014), until the predictable invasion of Ukraine by Russia in 2022.

Domestically, the Labour government presided over a period of economic growth and stability, and a vast increase in social support for families, low-paid families in work and education, reversing the trends of the previous Conservative administration and bringing public spending broadly in line with international standards (Ruth Lupton et al., 2013). The Labour government was also characterised by further neo-liberalisation of the political economy (Fuller and Geddes, 2008) and the welfare state, with increasingly targeted and limited benefits (Faucher-King and Le Galés, 2010), and significant investment in public infrastructure, often financed by public-private partnerships which proved to be a costly form of finance (Sclar, 2015; Musson, 2010).

The Labour government had a specific focus on education and social mobility, and support for young people and families, with the 1997 election manifesto mantra "education, education, education". There were a variety of initiatives to encourage local and national collaboration to enhance education in poor-performing areas and those with disadvantaged populations. Spending on schools was increased by 35.5% (Sibieta, 2021). 'Inclusion' was the stated policy, whereby young people with labels of SEND were increasingly educated within mainstream schools. For example, in 2007, 57.3% of young people with 'Statements' (the highest level of support) of Special Educational Needs (SEN) to use the terminology of the 1990s and 2000s attended mainstream schools (Department for Education, 2015). The majority of young people with SEN were educated in mainstream schools, as most young people identified as having SEN do not receive this highest level of support.

However, even in the heady days of the height of inclusion, the fact remains that in 2007, 37.9% of young people in the UK with Statements of SEN attended special schools (Department for Education, 2007, 2015). This statistic demonstrates, as reiterated in Chapter 8, that SEN policy was never fully inclusive of the whole diversity of mind-body-emotional characteristics. It was also evident that this shift was not fully resourced, with young people often being educated in mini-institutions and segregated within mainstream schools (Holt, 2004). Nonetheless, it is important to note that this research for the most part took place within a context in which resources for education were much higher than they are currently and in a political climate when state education, childhood and families, and the inclusion of young people with SEN were a fiscal and political priority.

The context has changed dramatically since 2010, precipitated in part by the global financial crash of 2007 and 2008, but also by a self-conscious and reflexive revanchist politics of Austerity in the UK, Europe and many nations of the Global South. In this context, Hastings et al. (2015: 1) emphasise:

> Local authorities in England lost 27 per cent of their spending power between 2010/11 and 2015/16 in real terms. Some services, such as planning and 'supporting people' (discretionary social care with a preventative or enabling focus) have seen cumulative cuts to the order of 45 per cent.

This is a disinvestment which disproportionately impacts upon less affluent people and places (see also Horton and Pimlott-Wilson, 2021; Katz, 2011).[3] Given the timing of the research in this book, it is possible to witness a shift from the relatively abundant, if neoliberal, public sector of the Labour governments of the late 1990s and 2000s to the Austerity cuts that began with the Conservative-Liberal Coalition in 2010.

The research presented in this book took place across a variety of school and local contexts. The young people had a range of mind-body-emotional characteristics and subjectivities, although those with SEN(D) or experiences of mind-body-emotional differences are over-represented, with more than half of the young people who participated in the research having such labels. Research was conducted with young people aged between 7 and 16, and across a diverse range of types of school: mainstream schools, an ability selective school, mainstream schools with more formal (i.e., Local Authority Designated) and more informal (i.e., school organised) special units, special schools and a Pupil Referral Unit. All the participating schools were state-funded. The purpose of the book is not to identify which type of school is best for young people with labels of SEND; indeed, when explicitly addressing that question there is no definitive answer (Holt et al., 2019a). Rather, there are practices and the deployment of "powers and resources" (Philo and Parr, 2000) at a variety of educational scales which promote opportunity, equity, inclusion, and appropriate support and education for young people, and those which do not. These contexts are much less about whether schools are 'special' or 'mainstream' or have dedicated units: it is about the deployment of 'powers and resources' and how they provide young people with opportunities to connect to other young people in a respectful and equal way, and whether they provide scope to develop social and cultural capital and for educational success. Promoting and supporting access to a fulfilling and appropriate curriculum is important, as is providing opportunities for social encounters. These are all contextualised by the socio-spatial context of the schools and the connections within and beyond the school spaces, which are afforded by young people's social connections; these can differ greatly according to location, as is further explored in Chapter 8.

2.6 Limitations to voices, accounts and empowerment, and the importance of young people's perspectives

It is equally important here to reflect upon the difficulties and challenges inherent in trying to prioritise the experiences of young people, and to admit, as Facca et al. (2020) emphasise, that the words and pictures on the page are not directly representative of young people's voices but have emerged from intergenerational engagement with the young people and adult researchers. As *contextual bodies/subjectivities/agencies*, young people's (and indeed our own) experiences are not fully knowable (see also Fielding, 2004). This presents a challenge, given that it is also important to take seriously and listen to young people's experiences and knowledge, which are so often sidelined. In the book, I have endeavoured to work with this tension, by representing young people's words and works as presented to me. Ethnographic observation helps to compare their accounts to my/Jennifer's situated and partial observation and to ask questions where these differ. Research with adults, teachers and parents primarily, helps to give a perspective of broader contexts and can be used to situate and contextualise some of the young people's experiences, as they might know more about certain contextual matters. The one thing they certainly do not know more about is young people's own experiences of the world. Young people are knowledgeable about their own lives. They are reflective and often rational, and their perspectives matter, even if they might be partial and situated, as, indeed, are all our perspectives and reflected experiences. These questions lead me on to discuss agency, subjectivity and power in the following chapter.

Notes

1 I develop Parr and Butler's (1999) concept of mind-body differences to also add the emotional, to emphasise social and emotional differences. These perspectives emphasise the corporeality of differences, how they intersect with and are forged within socio-spatial contexts, the interconnection between mind-body-emotional states, how people can be disabled via social and emotional differences which may not be tied to specific impairments or identified conditions, and that dis/ability is a continuum rather than a dualism between the disabled and non-disabled, as is commonly presented.

2 All of the names of participants are pseudonyms.

3 The level of revanchism has intensified in openness and intent since 2018 and the landslide election of the Conservatives, with Boris Johnson as a figurehead of the most right-leaning government the UK has witnessed since the francise was extended to the general population, at least. These events are, of course, tied to shifts in other 'democracies' which have been threatened and destabilised by populist governments (often with links to Russia's Putin – see for instance Intelligence and Security Committee of Parliament (2020) on Russian influence on UK politics) that appeal to soundbite politics (and probably invest huge sums in psychological research to identify the most pervasive soundbite, possibly funded by Putin and Russia's kleptocracy), such as Bolsonaro in Brazil and of course Trump in the US. Trump and Bolsonaro's attempted insurgencies in the face of their subsequent electoral defeats demonstrate the fascist and autocratic tendencies of, perhaps, a significant minority of their supporters. The disinvestment in social reproduction and providing a liveable life for groups of vulnerable people, such as disabled people, stands in sickening contrast to the profligate waste of funds through incompetence during both the Covid-19 pandemic and the current policy of subsidising fuel bills with public funds whilst energy companies make record profits. The first point is evidenced by the Public Accounts Committee report which states:

> Government has risked and lost "unacceptable" billions of taxpayers' money in its Covid response – and must account to the generations that will pay for it. (Hillier, Public Accounts Committee, 2022)

3 Young people's friendships and power

In this chapter, I set out why young people's social relationships matter: how and why they are powerful. In addition, I clarify how young people's agencies are theorised in the book, which I pull together within the concept of young people's *contextual bodies/subjectivities/agencies*. Rather than focus on young people as individual coherent wholes, I develop the idea of young people as dynamic, porous and connected bodies/subjectivities/agencies. Drawing upon a critical yet engaged relationship I have had with Non-Representational Theory and feminist post-structural theories, alongside a political commitment to young people's agencies, the idea of young people as *contextual bodies/subjectivities/agencies* demonstrates how they are becoming and emergent, how they become specifically within particular constellations of power, dynamic and yet also powerful agencies with unique personalities and perspectives and the power to reflect, affect change and, consciously and beyond consciously, reproduce or challenge enduring 'axes of power' (class, gender, race, dis/ability, religion and so on), alongside more subtle connections and differentiations (Cockayne et al., 2020). The political and conceptual adherence to the importance of young people's (even very young people's) agencies stands as a corrective to much educational research, policy and practice.

Whilst schools are often viewed as important spaces for transforming societies by developing young people in specific ways, young people's own role in this, their agencies within this process, are often ignored. In this chapter, I emphasise that the tendency to sideline the considerable agencies of young people within schools (and societies more broadly) must be critically rebuked because young people's social relationships are powerful. In schools, as elsewhere, young people make choices about who they socialise with and who they might exclude or marginalise in more and less subtle ways. These social performances are conscious acts of friendship, but they do more than just identify friends and those who are not friends. These performances identify who is like me and who is not like me, and they can reaffirm or contest expected ways of being tied to gender, class, dis/ability, race/ethnicity, sexuality, religion and subtle differences which might only exist fleetingly and/or in schools.

I draw upon the canon of geographical and social studies of children and young people. These have foregrounded young people's own agencies – the importance

DOI: 10.4324/9781003028161-3

of engaging with young people themselves – and their actions and their reflections, in an important political, ethical and scholarly move, and in contrast to the way young people often continue to be positioned as objects of socialisation and education. This is a critical development, and more can and should be done to engage educational scholarship and policy with young people themselves. Such engagement continues to be limited, and often tokenistic, despite the rights to participation enshrined within the United Nations Convention on the Rights of the Child (United Nations, 1989).

I go on to position the importance of young people's own social relationships within the concept of embodied and emotional social capital. This begins to reflect upon both the centrality of young people's sociality and the power embedded in social relationships and connections. The remainder of the chapter delves deeper into what is happening in these social relationships and critiques the self-evident and transparent idea of the power of young people's agencies. Critical to my examination of young people's agencies and social relationships is an overarching critical understanding of agencies, both in general and specifically in relation to young people. People are not all-knowing but are contextualised, constrained (and enabled) in specific ways. Young people's agencies and subjectivities are specific, given the bodily, cognitive and emotional development of children and youth. Young people's bodies-minds-emotions are specifically, though certainly not uniquely, dynamic. Young people's specific dynamism has implications across the life course. Our childhoods forge our adult bodies-minds-emotions, such that children and youth embody a "prefigurative politics" (Jeffrey and Dyson, 2021; see also Jeffrey, 2013) where social experimentation forges the actual material bodies and minds of young people in ways that embody future aspirations for societal transformation.

In the remainder of the chapter, I draw upon four key interconnected interventions which develop the central idea of young people as embodied and becoming subjectivities/agencies, and which are summarised by the theorisation of young people as *contextual bodies/subjectivities/agencies*. The first idea is the embodied nature of young people's agencies. The second is the contextual and dynamic nature of young people who *become* differently in different social, spatial, historical, political, economic and cultural contexts. The third is understanding young people as nodes of the intergenerational reproduction of enduring differences. Fourth, I emphasise the powerful nature of young people and specifically their socialities, and their power to reproduce but also to challenge and change enduring broader-scale inequalities via their everyday performances.

The *material, corporeal bodies* of young people matter and are matter. It cannot be denied that a baby's body is substantially different from that of a five-year-old and an eighteen-year-old. Of course, there is vast diversity within age ranges. Bodies are not pre- or asocial; rather, they emerge within specific frameworks and contexts, both in terms of food availability and type, regimes of exercise and sleep, clothes, footwear and so on and also in terms of norms of bodily ability and disability, perceptions of what bodies can and should do. Material bodies are contextual and are forged and become specifically in a dialogue between 'nature' (materiality, genetics, etc.) and 'nurture'. Recent studies of epigenetics have demonstrated how political and environmental contexts forge bodies in ways that are passed through generations: environmental factors, such as pollutants, can change genetic coding (Guthman and Mansfield, 2013). I adopt and adapt Bourdieu's concept of habitus as a useful heuristic device to conceptualise the intersections between social contexts (which are precarious and generated anew, despite a seeming fixity and despite many regularities of enduring inequalities), and dynamic, porous, interconnected bodies.

Young people's embodied being and becoming subjectivities/agencies are *contextual.* Young people become in material and social senses differently according to, quite simply, where and how they grow up. Social and spatial contexts forge and make young people in specific ways, and the limits and potentials of these contexts forge young people's subjectivities in ways which constrain and frame their agencies – limiting what is seen as possible and their potentialities in ways which might be conscious and known but are also beyond conscious and habitual. Young people as beings and becomings are embodied and the material bodies-minds-emotions of young people matter. These bodies are interconnected to the social; they are forged in specific ways in dialogue with the contexts of their emergence.

Young people are nodes of the reproduction of enduring differences. This is a process which I shall call, drawing upon Judith Butler (Butler, 1997), *subjection.* The contexts of the emergence can be changed. Importantly, the potentials of the agencies of young people are always constrained within powerful frameworks which are often not obvious to the young people themselves. Their everyday socialities reproduce enduring, intersecting, axes of power relations, such as gender, class, dis/ability, race and sexuality. They also have the power to change and transform these enduring axes of power relations. These potentials are influenced by the broader social and spatial contexts of young people's emergence, as established earlier.

Finally, young people's social relationships are *powerful.* Through processes of subjection, some of the powers of young people's friendships are working at a subconscious level through practices and performances that are not fully conscious or reflexive (as well as those that are conscious and deliberate). These serve to reproduce and/or have the potential to challenge enduring subject positionings, as well as more subtle connections and differentiations (Cockayne et al., 2020) which are not sedimented into societally pervasive "axes of power relations" (Butler, 1990). These performances have powerful impacts on young people's experiences of all elements of schools, including more formal aspects (and vice versa). This interrelationship can reaffirm and contribute to reproducing education inequalities around socio-economic class, race/ethnicity, gender, dis/ability and their intersections. Of most interest is, perhaps, the *power* of young people to *challenge and change* enduring axes of difference, producing new ways of being and new connections or lines of flight. This is developed further in the subsequent chapter, where I examine immersive geographies.

3.1 Young people's agency – a call to critical educators

Since James et al. (1998) called for the new social studies of childhood to take seriously young people as critical social actors, also reflected in Holloway and Valentine's (2000a) seminal edited collection, the fields of social and geographical studies of childhood and youth have bloomed with detailed, empirical accounts of young people as social agents. The central pillars of geographies and social studies of young people – that young people are social agents, that childhood is a social construction and the importance of young people's participation in research – remain as salient today as they were in 1998 (James et al., 1998; see also Holloway et al., 2019):

> that children could – and should – be regarded as social actors, second, that childhood, as a biological moment in the life course, should nonetheless be understood as a social construction; and finally, there was methodological agreement about the need to access children's views first hand.
>
> (James, 2010: 216)

These central tenets have raised important questions for researching young people and have transformed how children and young people are understood, at least within the fields of social and geographical studies of young people. A central orthodoxy is the need to research *with* young people, given adults as proxies cannot understand young people's interpretations of the world around them, by engaging reflexively with young people's own experiences. The field abounds with scholarly accounts of young people's experiences and interpretations of a vast array of social and spatial contexts across the globe. For instance, in a single issue of the journal *Children's Geographies* (Volume 20 Issue 2), we encounter young people's experiences of urban, island, school and preschool spaces and children from younger than four years to older youths, in a range of national contexts from Greece to Sweden, Ghana and Malaysia. We encounter young people in a variety of social circumstances, including precarious migrants. Almost without exception the voices and experiences of young people are prioritised. The Springer collection of *Geographies of Children and Young People*, brought together by Tracey Skelton as the editor in chief (Skelton, 2016), consists of 12 volumes, each of around 20 chapters, relating to research with young people. Scholarship ranges from arguably more expected topics, such as *Place, Space and Environment* (Nairn et al., 2016), and *Play and Recreation* (Evans et al., 2016), to those which challenge the way young people are understood, which engage with *Politics, Citizenship and Rights* (Kallio et al., 2016) and *Conflict, Violence and Peace* (Harker et al., 2017). These are selected examples, which point to the importance and depth of geographies and social studies of children and young people, within which the agencies of young people remain paramount despite some challenges and calls to move "beyond agency" (Kraftl, 2013).

One commentator who has critiqued the lack of an engaged theoretical evaluation of young people's agencies is Prout (2000). Alongside suggesting a more nuanced and critical notion of agency which engages with development and bodies, he argues that the concept of young people as "social actors has 'won wide agreement'" (p. 2). I disagree with this assessment; although within social and geographical studies (and here I include anthropological approaches, sociologies of education and so on) *with* young people, there is broad agreement that young people have agency, such an approach is not widespread, either within the more broadly aligned disciplines (e.g., Geography and Sociology) or within the disciplinary approaches that have the most impact upon young people's lives, such as education and social work. Perhaps partly as a result of some of the critiques Prout levelled at the field, such as a tendency not to engage fully with other approaches to childhood and youth, in schools, in social services and in other fields that most directly affect young people, the central mantra of children and youth as social agents has not been adopted wholeheartedly. Curricula continue to be imposed on children and youth (and teachers), rather than being part of an engaged process that connects with young people's perspectives (see Olsson, 2009).

Given the vast richness and depth of the fields of geographical and social studies of young people, it is problematic that these fields have had less influence on the arenas of young people's everyday lives than we might have hoped. In advocating a more thorough engagement with young people's agencies in schools, my arguments align with critical pedagogic work, and Aitken's comments resonate:

> Is this what I want for the children with whom I connect? A sense of wonder, enchantment and engagement with life rather than the pressures foisted upon them from a neoliberal structure that not only forecloses political potential but sucks out life and passion in the name of efficiency and rationality.
>
> (Aitken, in Kohan et al., 2015: 407)

In the UK as in the US and many countries across the globalised world, we have not only 'sucked out life and passion' from education but failed to acknowledge and take account of young people's own creative agencies. This tendency is exacerbated within neoliberal shifts towards high performance and selection within knowledge economies, which enshrine the 'normally developing child' alongside instrumentalist pedagogies.

As I emphasised in the first chapter, my research has engaged with young people (and adults) with a variety of mind-body-emotional characteristics, with and without labels of Special Educational Needs and Disabilities (SEND) and from a range of socio-economic, racial, ethnic, religious and geographical backgrounds, in English schools. Within the research, it has become apparent that young people's own social relationships are foundational to their experiences of school in ways which are often overlooked by adults. Young people who had friends and good friends broadly liked all elements of school, and those who were left out, bullied, excluded or marginalised disliked school – including formal learning elements. This is not a simple one-way and linear relationship, since many of the young people who were left out, bullied, marginalised and/or excluded in young people's relationships also had negative experiences within formal elements of the school, and the two were interconnected. Indeed, the way young people are represented and performed as 'other' through adult practices and the material spatialities of the school spaces are an important context in which young people learn how to perform their subjectivities (see Chapter 8).

Young people's own relationships were pivotal to their experiences of school and important to their level of engagement; yet, although teachers did reflect on young people's social relationships and building of social and emotional skills, endeavours to do this were not pivotal or foundational to the running of schools and classrooms, but an additional consideration. Consequently, I call on educators, policy makers, parents and others to appreciate that young people's social relationships are powerful. In the following section I outline my conceptualisation of embodied emotional and social capital to capture some of the powers invested in young people's social relationships.

3.2 Embodied social and emotional capital

3.2.1 Embodied social capital

Social capital continues to be an important concept in policy forums, and it broadly emphasises the importance of social networks and connections to individuals and groups. It is a useful concept for this book, which seeks to emphasise the importance of young people's own sociality. Social capital, or the idea that 'it's not what you know it's who you know' is a pervasive and important idea in policy and academic studies. The importance and powerfulness of social relationships are broadly agreed, and yet the pervasiveness of young people's own sociality to their experiences of school is largely overlooked or relegated to peer affects. Although there is a broad acceptance that 'who you know influences where you go', the theories of social capital are highly disputed. Similarly, whilst policy makers and politicians find social capital to be a powerful tool, many critical academics have negated the powers of the concept. This contrast is largely due to the claiming of the field of social capital by conservative commentators, notably Robert Putnam, who posits social capital as causal and a general social good. Pierre Bourdieu's theories of social capital are distinct and specific, and critically examine social capital as a mechanism for reproducing inequality and dis/advantage. The concept of social capital is often dismissed by critical scholars, and this is partly due to the marginalisation of Bourdieu's accounts compared to those of Putnam and to a lesser extent Coleman (Coleman, 1987; see also Schaefer-McDaniel, 2004).

Putnam's version of social capital has been hugely influential because it highlights the importance of social relationships to all aspects of life. This point I accept. Beyond this I reject entirely Putnam's theorisations, which rely on a spurious statistical link between social relationships, civic engagement and all kinds of social good (see Fine, 2002). Indeed, Putnam problematically provides social capital with a spurious causality which in effect blames poverty on poor people's lack of civic engagement (Putnam, 2000). This simplistic quick-fix approach is of understandable attraction to policy makers and shapers, specifically because it does not require any change in the structural order of societies. Critical scholars, such as Das (2004: 27), lambast Putnam's theory of social capital:

> it is untenable to posit social capital as an independent variable and poverty as a dependent variable because the economic-political conditions of poor people have an enormous constraining effect on social capital itself and its supposed material benefits for the poor.

Putnam's view of social capital as both an independent and causal variable is the opposite of critical understandings of social capital emerging from Bourdieu (especially 2018). Bourdieu's theories of social capital have been much less influential than his accompanying (and admittedly more clearly conceptualised) theories of cultural capital in critical educational sociology, geography and social studies more broadly. Bourdieu theorises social capital as a specific form of capital tied to social relationships and networks:

> the aggregate of the actual or potential resources which are linked to possession of a durable network of more or less institutionalized relationships of mutual acquaintance and recognition – or in other words, to membership in a group – which provides each of its members with the backing of the collectively-owned capital, a 'credential' which entitles them to credit, in the various senses of the word.
>
> (Bourdieu, 2018: 249–250)

Bourdieu emphasises the interconnections between social capital and other forms of capital – social capital does not operate in isolation; it is not some kind of sovereign capital which can transform all the social and economic ills of society. Rather social capital interacts with economic and cultural capital.

Indeed, Bourdieu is *most concerned* to emphasise how social and cultural capital serve to reproduce enduring intergenerational privileges and disadvantage. Bourdieu links social capital directly to economic capital and enduring socio-economic inequalities. This may be argued to be a form of economic reductionism; however, it also reinstates the importance of material advantages and inequalities. These were apparent in 2008 and were one of my motivations for writing about embodied social capital; however, in the wake of the global financial crisis which affected the Global North along with the crises previously shaking countries of the Global South or the Majority World (Harvey, 2011), the rise of populism and intensification of neoliberalism, material dis/advantages and inequalities are becoming ever more entrenched, insidious and aggressive (Dorling, 2018; Katz, 2018). Indeed, the poor are viewed as moral failures even to themselves (Shildrick and MacDonald, 2013) or criminalised (Wacquant, 2009).

Schools, including state schools, for Bourdieu (e.g., Bourdieu and Passeron, 1979, 1990) are a key site of the reproduction of privilege and disadvantage. Nonetheless, whilst Bourdieu's own studies of schools and other educational institutions as sites of the reproduction of inequalities have been pervasively influential

(Reay, 2004a), studies of the social capital of *young people* have been relatively marginalised (see also Schaefer-McDaniel, 2004). An important exception is Morrow's (1999, 2001) fascinating work on young people's social capital, health and well-being, which is limited only by the lack of awareness of the pervasiveness of the importance of young people's social and emotional capital to all aspects of their lives.

Bourdieu's notion of social capital was relatively untheorised in comparison to his broader canon, and to more thoroughly flesh out Bourdieu's conception of social capital, it is necessary to draw upon his broader work and to make some inferences. For instance, the aforementioned quote suggests that all members of a network have equal access to the social capital of that network, which cannot be the case. Individuals will be differentially positioned within that network, of course. Nonetheless, we can infer that Bourdieu's conception of social capital is a way of articulating both the importance of social relationships *per se* and the ways in which social relationships forge networks between people and connect them to other forms of capital so that advantages and disadvantages are further entrenched.

Being social and living in relation with others is critical, and Judith Butler (2004a) also emphasises the critical importance of social, psychic and practical interdependence and interconnections, the being in relation to others, through which social subjects are forged. Consider the difference in the access to other capitals gained by a popular boy at a highly prestigious fee-paying school such as Eton, compared to a marginalised and isolated young person in a comprehensive school, labelled as 'inadequate' by Ofsted, in a disadvantaged neighbourhood. Social relationships (or a lack of them) are critical in themselves.[1] Further, different social networks and people's different positions within these networks provide connections with other forms of capital – cultural in all its forms, as well as economic. There is a clear link between social, cultural and economic capital. Economic capital opens up spaces within which particular social relationships emerge (such as top-flight universities or elite schools). Cultural capital (such as educational qualifications, an appropriate set of demeanours and knowing how to behave) allow the "alchemy of consecration" (Bourdieu, 2018: 251) to develop relationships of trust, reciprocity and mutual obligation. Both social and cultural capital can be connected to economic capital – for example, knowing the right people to get a job or having high levels of qualifications and knowledge of the correct way of being in the world to succeed in any given field.

3.2.2 Emotional capital, recognition and a foundational interdependency

The concept of emotional capital nuances Bourdieu's arguments, to do two things: first, to emphasise the importance of emotional reciprocity and interdependence to sociality; second, to radically reconstitute the subject as never fully formed, always becoming and becoming in contextual relationship with others. Bourdieu's comments on social capital appear to imagine a rational, economically motivated actor whose primary objective in life is to use every mechanism at their disposal to enhance and/or maintain their privilege (cf., importantly, Bourdieu's ideas of habitus and praxis, discussed in part in the following). This, of course, may be the way people act and the source of their motivations at times, or what motivates some people all or most of the time. By contrast, however, most of our motivations are more complex and nuanced and governed by emotion, habituation and our beyond-conscious, alongside our strategic and rational, reflection. In line with an increase in interest in emotions and affect in geography and social sciences (as indicated by the journal *Emotions, Space and Society*, which was established in 2008) and a long history of critique of the rational economic actor theory of agency, with colleagues (Sophie Bowlby and Jennifer Lea) I have further developed the concept of emotional capital, first coined by Nowotny (1981) and further developed by Reay (2004b). Reay (ibid.) points out that Bourdieu underplays the importance of emotions in her work on mothers' emotional investment in their children's education; she proposes that emotional capital is another form of capital alongside the economic, cultural and social forms of capital which Bourdieu developed.

In positing the emotional as another form of capital, Reay underplays the significance of the emotions; instead, it is possible to view emotional capital as foundational to all aspects of life, given the fundamentality of emotions to social life (Bondi et al., 2007), and as a need to live within relationships of emotional recognition and interdependency is central to most humans' experiences. Emphasising the need for emotional recognition also highlights the corporeal and the interconnectedness of minds-bodies.

Reflecting on emotional capital emphasises the relationality of people to each other and (to a lesser extent) to non-human others, and the ways in which people become in specific contexts; Judith Butler (2004a, 2004b) emphasises the importance of physical and emotional interdependence: "we are, from the start, given over to the other . . . we are, even prior to individualisation itself, and by way of our embodiment, given over to an other" (Butler, 2004a: 23). This position of being 'given over', of being dependent physically and emotionally, upon others means that, for the most part, humans emerge as subjects within a vital interconnection with others. Dependent on others for bodily survival, humans are also dependent on others for a liveable life with emotional and social 'recognition'. The importance of interdependence to the emergence of the embodied subject/agent emphasises the ways in which power is constitutive of our subjectivities from the outset. This situation of the emergence of the subject in a foundational interrelationship with others helps to illuminate the ways in which humans *become* recognisable to themselves and others as subjects within constellations of power. This position resonates with psychoanalytic geographies (Kingsbury and Pile, 2014; Philo and Parr, 2003).

The (young) person does not precede the social encounter but becomes specifically within the context of the psycho-social interactions and the historically embodied previous interactions that they have previously encountered, in dynamic and generative formation with their corporeality. This idea of an inherent intersubjectivity and emotional interdependency as foundational to becoming a person or a subject questions the entire suggestion of a coherent and formed agency, as further elaborated in the following.

3.3 Young people as contextual bodies/subjectivities/agencies

Without moving away from the importance of engaging with young people's own social agencies, and young people as social actors, in a context in which this agency is continually denied, it is problematic to claim an unfettered social agency as a 'mantra' (Punch and Tisdall, 2012) without fully interrogating what is meant by agency or being a social actor. Indeed, as Sarah Holloway, Sarah Mills and I point out (Holloway, et al., 2019), it is problematic that agency was being claimed for young people as critical social and geographical studies of young people emerged alongside critiques of transparent notions of agency or indeed of any specifically *human* agency at all. The reality of the cognitive, biological and social development of young people as they age also abuts conceptions of young people which foregrounds their capabilities and denies how these capabilities are emergent and situated within changing dynamic bodies. This also impedes dialogue between critical social and geographical studies of childhood and youth and the fields which continue to have the most sway over young people's lives, such as education, social work, public health and medicine. Allan Prout (2000) pointed to this danger, which has not diminished or been overcome.

Sarah Holloway, Sarah Mills and I moved forward conceptions of young people's agency in our paper published in 2019, by developing the concept of young people as 'biosocial beings and becomings'. In this paper we maintained the central theoretical and political importance of foregrounding young people's agencies, given that they are so often overlooked as critical social actors in all kinds of arenas which most concern them. It was important to us, and continues to be important to me, that young people have their competences as critical social agents foregrounded, given that they are so often sidelined. Nonetheless, it is problematic to cast young people as unfettered social actors at the very same time as critiquing a notion of liberal agency in what we labelled post-structuralist feminist scholarship. Indeed, as Ruddick (2007) points out, the very notions of liberal agency that we claim for young people are the same conceptions of independent action by which all groups of people whose dependency cannot be concealed are cast as non-agents. In addition, of course, casting young people as unfettered agents is inevitably going to come up against their limitations as sovereign actors – young people are not all-knowing and all-seeing; this is a limitation of any claim of sovereign or independent agency. Yet, the agencies of young people are, specifically, embodied within dynamic, growing, changing bodies. Whilst I argue infants have agency, it is differently expressed than an adults' agency. Acknowledging infants' agencies challenges the very notion of what agency is (Holt, 2013; Holt and Philo, 2023).

So, here I would like to go further than Holloway, et al. (2019) to claim for young people's *agencies*, but a diffuse and connected sense of agencies, which is not tied to a specific and bounded *agent*, and one which is always constrained and enabled in conscious and beyond-conscious ways. These agencies are powerful, but some of this power is about reproducing (or potentially challenging) enduring and interconnected "axes of power relations" (Butler, 1990) in conscious and deliberate and in unconscious and non-reflective ways. I want to shift the emphasis away from young people as *beings*, which suggests a cohesive self; instead, I suggest an alternative: young people as dynamic ***contextual bodies/subjectivities/agencies***. This emphasises the porous and emergent nature of young people and their connections to other people and things in their becoming. Importantly, however, it does not deny an interior self, and in this I stand in stark contrast to some elements of Non-Representational Theory, and particularly in some recent accounts of the concept of encounter (Wilson, 2014; Cockayne et al., 2020), despite taking inspiration from these.

3.3.1 Embodied subjectivities

3.3.1.1 Bodies – matter, habitus, life course

Young people's bodies are matter, which does matter (Aitken, 2001; Prout, 2000; Hörschelmann, and Colls, 2009), and young people are lively, affective, emotional porous and connected bodies. Bodies-minds and emotional states are interconnected (see Parr and Ruth Butler, 1999; Hall and Wilton, 2017). Young people's encounters involve various intra-embodiments (Deborah Lupton, 2013),[2] and in the contemporary context, their encounters are usually hybrid, connecting far-flung places through social media and digital technologies, to material co-present moments in taken-for-granted ways which defy adult rationalities. Bodies are porous and connected, rather than bounded, and social relationships are tied to various interconnections between material bodies – where skin touches skin, not bounded but in the process exchanging microscopic particles which contain DNA, the very map of the individual. As Cockayne et al. (2020) point out, breath is shared and the air moves between subjects who are within close co-proximity – an innocuous idea when written, no doubt, but this very porosity of bodies and the sharing and circulation of animate and inanimate others (and the Covid-19 virus which hovers at the horizons of the animate and inanimate) through touch and respiration became the very reason why bodily co-presence was suspended and then limited for months in the UK and across the globe.

Young people's bodies are simultaneously the site of experience and a site of interpretation, in what Grosz (1994) has labelled "social tattooing". Importantly, these two elements intersect and intertwine. Through processes of normalisation (Foucault, 2003) and self-regulation (Foucault, 1977), performance and subjection, we regulate our bodies and embodied ways of being in the world within the context of the limits and potentials that others ascribe to us. We can self-consciously 'exceed' the 'exegeses' of power (Butler, 1997), although we can never fully escape them, because they are foundational to our coming into being as subjects. Our bodies-mind-emotions are forged differently through regimes of learning, diet, exercise and so on as these intersect with matter and discipline bodies into being socially appropriate.

Young people's bodies are specifically dynamic, developing and changing; yet importantly, this change is not universal, but is socio-spatially contextualised and, indeed, individual. This point is critical, because despite this, most social and educational policy assumes a normative and universal development which casts as 'other' those who fall outside and typically below expected age-related developmental milestones. Material bodies and their capabilities also matter. So much of what young people are expected to do is normalised through models of development which take an average and turn it into a regulatory norm. Young people whose bodies/minds/emotions will not be regulated within these norms of expectations of learning, behaviour and emotional regulation feel an embodied sense of dislocation, of frustration. These differences are real and material and can be frustrating and challenging, yet the material differences only emerge as such within the context of normative expectations.

Both Bourdieu's theory of habitus and Butler's theories of subjection, power and performativity help me to reflect upon how these social and spatial relationships forge bodies/subjectivities/agencies. Some scholars have emphasised the differences between Bourdieu and Butler's conceptions of habitus and performativity, and they do draw on differing philosophies. Bourdieu is viewed as more of a historical materialist, foregrounding the political economy, whereas Butler has been regarded as more of an idealist, foregrounding the power of cultural representation and diminishing the continued importance of economic inequalities (Lovell, 2000). I argue that in part because of these critiques, yet also because their work and perspectives do have points of connection and similarity, connecting Butler and Bourdieu provides a more complete picture of the importance of both socio-economic axes of inequality (which is forged on and through bodies) and those tied to other embodied identity positionings, such as gender, sexuality, dis/ability, race/ethnicity and so on. These aspects of power and inequality, advantage and disadvantage clearly intersect.

Bourdieu's theories of habitus have been widely influential within sociologies of education and geography, yet they have been interpreted in different and sometimes seemingly incompatible ways (see for instance Bridge, 2006; Reay, 2004a, 2004b; Waters, 2006; Darren Smith and Phillips, 2001). This is not surprising, given that Bourdieu himself conceptualises habitus in many different and sometimes competing ways (Lizardo, 2004). For me, habitus explains the ways in which bodies (including minds and emotional states) are porous, connected to and forged within relation to specific material, social, spatial and political contexts. Bodies-minds-emotions are a kind of sedimented, material history of our trajectory through space and time. Bourdieu gives an insight into how habitus is forged through the interconnection of people with 'environments':

> The structures constitutive of a particular type of environment (e.g. the material conditions of existence characteristic of a class condition) produce habitus, systems of durable, transposable dispositions, structured structures predisposed to function as structuring structures, that is, as principles of the generation and structuring of practices and representations which can be objectively "regulated" and "regular" without in any way being the product of obedience to rules, objectively adapted to their goals without presupposing

> a conscious aiming at ends or an express mastery of the operations necessary to attain them and, being all this, collectively orchestrated without being the product of the orchestrating action of a conductor.
>
> (Bourdieu 1977: 72)

In his later work (e.g., Bourdieu, 1984, 1990, 2020), Bourdieu refines the somewhat fixed view of the 'outside' reality that 'structures' habitus. At the time when Bourdieu was devising the concept of habitus, this was a nuanced attempt at rethinking the dualistic notions of structure/agency and body/society to emphasise that material mind-bodies emerge within specific social, economic, cultural and political contexts. The external structures do not need to be viewed as 'fixed' but are themselves re-enacted and remade through everyday practices and are subject to change. Butler's (1999) critical engagement with habitus then is slightly acerbic; actually, their projects are rather similar – albeit that Butler does not foreground (and indeed largely overlooks) the material advantages and disadvantages tied to socio-economic conditions or 'class'.

In a paper with Sophie Bowlby and Jennifer Lea, we set out our view of habitus:

> Habitus is a set of embodied dispositions – tastes, preferences, ways of being, accents, and so on, which form an unconscious backdrop to (future) social encounters (Bourdieu and Thompson, 1991; Reay, 2004a). As bodies can be conceived as porous and unbounded, connected to other bodies and only becoming in specific spatial contexts and in relation to a variety of human and non-human actors (Colls, 2012), habitus can be understood as simultaneously a property of individuals and collectives, and has even been tied to particular spaces (e.g. Smith and Phillips, 2001). Thus, habitus mostly operates at a sub- or beyond-conscious level; transformation of habitus largely occurs via beyond conscious responses to new 'fields' rather than via deliberate attempts at change (Bourdieu and Thompson, 1991).
>
> (Holt et al., 2013: 34)

Importantly, habitus is the way in which the socio-spatial contexts of young people's emergence, including their social relationships and connections, intersect with porous, material bodies, to forge bodies in particular ways, providing a context for future socio-spatial encounters. Habitus is helpful to examine how young people's subjectivities are inherently dynamic and contextual, as indeed are Butler's theories of performativity and subjection. Habitus provides a useful device to also think through how young people's embodied subjectivities are not endlessly dynamic but also forged in relation to material corporealities and imprinted by the history of previous social contexts. The socio-spatial contexts that young people encounter in their childhood and youth intersect with their bodily matter to forge a habitus that continues throughout the life course, albeit that this might change and transform as it connects to new times and spaces; yet childhood is a particularly malleable period and bodies-minds-emotions are arguably laid down during childhood in ways which are more difficult to transform later in life – our childhoods and youth are always an embodied and psychic presence throughout our life course.

3.3.2 The contextual and dynamic nature of young people's agencies

Both habitus and performativity emphasise the contextual nature of young people's subjectivities/agencies. Butler (1997, 2004a) takes this forward in her discussions of subjection and recognition. Butler (2004a) emphasises that emotional and material interdependence is foundational to the formation of subjects, and thus people are never bounded but are always articulated in emotional and material relation to others. Critical to discussions of emotional and material interdependence is the central idea of recognition. Butler approaches the concept of recognition from a variety of perspectives, and for me the most influential are her discussions of Hegel and Jessica Benjamin's psychoanalytical theories. According to Butler (2004a), Hegelian notions of 'recognition' emphasise a constitutive search for recognition, which is foundational to the formation of the self. She argues that it is only through the recognition of ourselves by others that: "any of us becomes constituted as viable social beings" (Butler, 2004a: 2). Therefore, the formation of our understanding of ourselves as people is predicated on a recognition of us as people by others.

Exploring the first point, the emotional need to be recognised is central to the emergence of the subject and critical to how subjects emerge within power (cf. McNay, 2004). The need for emotional recognition sets a context for *why* people regulate their bodies according to the existing exegesis of power. It helps us to understand how power operates in ways that are generative and constitutive of subjects and agencies, alongside being constraining and limiting. As parents and educators we understand the soft, constitutive power through which we forge children and young people to be and become, to reach their potentials and to contribute to society. We might not be so fully aware of how we frame and constrain their horizons.

All people are arguably emergent and dynamic, *contextual bodies/subjectivities/agencies*; however, it is critical also to acknowledge that young people are specifically dynamic. The processes which forge subjectivities are laid down in enduring ways in childhood, arguably particularly earliest childhood (Butler, 2004a; Pile, 1996; Bourdieu and Thompson, 1991) in ways which *can* be challenged or changed in later life, but which are resistant to change, and perhaps experienced as 'natural'. Butler (1997, 2004a) points to the importance of infancy to generating an apparently interior or socially anterior psyche, which is socio-spatially constituted (Holt, 2013). Such a view has resonance with psychoanalytical geographies (e.g., Kingsbury and Pile, 2014; Thomas, 2005, 2010; Pile, 2010; Davidson and Parr, 2014; Aitken and Wingate, 1993; see Chapter 4).

Although habitus can change in relation to future encounters in new socio-spatial contexts, as an internal, unconscious or reflexive backdrop it is resistant to change – and this is precisely the point. Although forged within social contexts, and dynamic, bodies and minds are not endlessly dynamic. They have a memory or an imprint of where they have been before, they are embodied and have a resistance to change. Early childhood is particularly pertinent to the forging of habitus in ways that become embodied and can seem natural and have a resistance to change. For instance, we cannot undo our tastes and dispositions or our accents without a significant amount of conscious effort. Of course, the 'fields' or external realities within which habitus emerges are not themselves static, but can be challenged and changed. Habitus is a concept of space and time and helps to clarify the life course and intergenerational elements of childhood. Young people's bodies-minds-emotions are forged within specific contexts with access to a different set of capitals, resources, ideas and so on. Our childhood contexts are embodied and stay with us throughout our life course. Our families' socio-spatial context and access to capitals is an important context, and habitus, capitals and indeed much of Bourdieu's work provide insight into the intergenerational reproduction (and less so the transformation) of advantage and disadvantage.

Although habitus has been often drawn upon in sociologies of education to understand how 'external realities' forge young people's embodied subjectivities, the focus of these examinations has been largely on the ways habitus is forged in families, and often the focus is on adults, rather than children and youth themselves. How schools forge 'habitus', and an embodied subjectivity of young people, has been less often examined. However, given the amount of time young people spend in schools, they are important contexts for the emergence of habitus, and this is explored within this book. Further, although Bourdieu refines his early view of the 'external reality' to become more dynamic, sociological studies focusing on habitus tend to over-prioritise *reproduction* of education inequalities above the potential for transformation and change.

Although Bourdieu is open to other social differences, both he and the scholars who have followed him tend to focus on social class (Alanen et al., 2015). These tendencies are justifiable; now, 14 years on from first writing about embodied social capital, I am more convinced of this justification, given the changes wrought in the UK by 13 years of increasingly neoliberal Conservative rule, and globally by entrenched neoliberalism (Katz, 2018). First, education inequalities *are* often reproduced and less often transformed, so that through the embodied experiences of generations of young people, educational inequalities endure through time and space. Yet, there is the potential for transformation, and again, this book is keen to expose the ways in which transformation does happen in school spaces. Second, socio-economic differences and class remain pivotal to young people's opportunities and experiences, yet young people's subjectivities are intersected by a variety of axes of difference and education inequalities are also intersectional. In understanding this intersectionality, dynamism and potential for transformation (if perhaps slightly overemphasising this element), Judith Butler's theories of performativity and subjection help to remind us of the diverse operations of power through differing and intersecting axes of power – class, gender/sex, sexuality, race/ethnicity and, importantly, as extrapolated to dis/ability.

These arguments point to the fact that people's bodies, identities and subjectivities (and I am writing specifically here about young people) are inherently socio-spatially contextual. They come into being in specific spatial contexts. Young people become who they are in specific places, in specific moments in time and space, and a particular "thrown togetherness" (Massey, 2005) of spatial and historical contexts, and this forges who they are in important ways. Young people's social relations and how subjectivities are played out within them are critical in forging young people's own embodied subjectivities/identities, such that they can transform disability (for instance) into new 'lines of flight', new horizons, new potentialities and ways of being, as I will begin to account in the subsequent chapters.

These social relationships are themselves emerging within a context of all the past encounters of the players within these events (adults, young people, powers, resources, material spaces and things) and their connections (bodily, local, global). This is not to deny the role of matter or emotions, of genes or personalities, rather that the embodied person becomes through an iterative relationship between these materialities and the contexts of their emergence. There is much debate about 'interiority'; however, for me, as I discuss in the following, an interior reflective and thoughtful self is important. Drawing upon psychoanalytic geographies, I argue here that there *is* an interior psychic life, which, whilst it might not be pre-social, is certainly experienced as extra-social, and internal, and is an interiority that is brought to every social encounter. It is, at least, the interiorised and embodied processes of past encounters intersecting with materialities and personalities. This interiority *can* change, but it is not immediately pliable and dynamic. It is an interior mind or psyche, but this is inherently interconnected to bodies (Parr and Butler, 1999) as thoughts are affective and corporeal and feelings are also affective, corporeal and yet also cognitive, and can be rationalised and contained. In Chapter 8, I reflect more fully upon the connections between the small scales of inter-embodied (Deborah Lupton, 2013) socio-spatial practices of young people in schools and broader-scale socio-spatial processes – for now I will keep these implicit. People continue to be dynamic throughout their lives; however, I argue that young people are specifically (though not uniquely) dynamic.

Given the specificity of the dynamism of young people's bodies and minds, and the fact that they are invested with futurity, given also that schools are institutional spaces within which young people are contained, schools and the young people within them can be the focus of social experimentation. This includes bringing together 'different' groups for prolonged periods of time, which is not possible in other spaces, with an attempt to change the way different groups view each other in the future. Examples include Northern Ireland (Department for Education, 2022) and Waterhead Academy, Oldham (Edmonds, 2015). A further example is the 'inclusion' of young people with mind-body-emotional differences, and labels of Special Educational Needs and Disability into mainstream schools, which has been the focus of my research. Something about this specific period of growth and change and having lived less time in the world does present opportunities for young people to do differently the things that adults have got so very wrong. It does also seem unfair and unrealistic to ask young people to change all the mistakes of their forebears. David Cameron (2015) in his speech about extremism put integrated education front and centre of his strategy to counteract extremism:

> It cannot be right, for example, that people can grow up and go to school and hardly ever come into meaningful contact with people from other backgrounds and faiths. That doesn't foster a sense of shared belonging and understanding – it can drive people apart.
>
> (also referenced in Edmonds, 2015)

Young people can be viewed as embodied "anticipatory politics" (Jeffrey and Dyson, 2021) where education and other social experimentation are enacted through a biopolitics of forging specific young people's embodied subjectivities in an attempt to generate a particular future society.

3.3.3 Young people as nodes in the reproduction of enduring differences in space and time: performativity and subjection

The theory of habitus has many connections with Judith Butler's ideas of performativity and subjection, which conceptualise how people become within specific social and political (and less obviously economic) contexts; and indeed, Butler has critically engaged with Bourdieu's habitus (Butler, 1999). Butler's influential concept of performativity emphasises that material bodies become stylised into seemingly fixed categories (of gender and sex particularly) through repeated performances in time (Butler, 1990, 1993, 1999). Butler's work draws upon critical analysis of media sources, and as such, she has been accused of denying the material reality of bodies (see also Holt, 2013). Although this critique has some foundation, certainly in the ways in which Butler herself theorises bodies, others have been inspired by Butler's work to produce nuanced accounts dealing with socio-spatially constituted material bodies in specific performed spaces (e.g., Gregson and Rose, 2000).

In this book, I also want to consider Butler's work on subjection/subjectification, which is a Foucauldian notion. Here I label this subjection, although it is often labelled subjectification (as developed in the History of Sexuality series, Foucault, 1978, 1984a, 1984b; see Foucault, 1982; McNay, 1994). Subjection is a creative play of power in which the subject is constituted, which simultaneously brings the person into being and limits their possibilities. Subjects emerge within specific contexts of power – although these are never fixed but reiterated and dynamic, with the potential for change. The power here is more expansive than Bourdieu's understanding of capitals and defines all kinds of positionings. It also breaks the connection with the material political economy and economic advantages and disadvantages. This break is helpful in facilitating an analysis of broader operations of power. However, this break has also been part of a wider shift away from analysing socio-economic inequalities and advantages/disadvantages at precisely the same period in which Dorling (2018) and others emphasise that such inequalities were increasing. Nonetheless, by intersecting Butler and Bourdieu, I intend to draw upon this more expansive notion of power to reflect upon a broad range of intersecting power relations whilst continuing to raise awareness of entrenched and often increasing inequalities and disadvantages tied to capitals and the economic. Indeed, being on the more marginalised side of these key categorisations, such as gender, race/ethnicity and dis/ability, has consequences for access to capitals and material inequalities. These axes of power of course intersect. In my work I have most often focused on dis/ability, and this focus continues within this book, yet I am also interested in how dis/ability intersects with other axes of power relations, most particularly socio-economic dis/advantage, capitals and class.

Subjection is done by the subject to itself. People willingly, although often in ways that are not conscious, situate themselves within frameworks of power (which are, however, never complete and always reiterated and worked anew). This is not (only) an external enactment of power over individuals but a creative play of power within the constitution of the subject, which simultaneously limits the person's possibilities and brings the subject into being. Subjection allows people to enact agency, permitting them to 'exceed' (though not fully escape) the existing exegesis of power (Butler, 1997). Butler (2004a) emphasises that subjection is relational and occurs within the context of emotionally and physically interdependent relationships. Butler suggests that it is this pervasive human need for emotional *recognition*, alongside their material interdependence, which makes people vulnerable to subjection.

Notwithstanding their somewhat differing philosophical traditions, with Butler emphasising discourse and language and Bourdieu more closely aligned to (historical) materialism (McNay, 2004; Grenfell, 2004), from a pragmatic perspective (West, 1989), there are commonalities in the practical implications of Bourdieu and Butler. Both authors deconstruct the body/society and structure/agency dualisms, emphasising how agency emerges within the context of broader cultural social, political and economic conditions. These constrain, but do not fully determine, agencies and what a person can be – embodied beings and becomings internalise the conditions of their emergence, such that 'structures' do not exist externally to the person but become part of who they are. The broader 'structures' are not fixed or given, but are dynamic and shifting, either through deliberate conscious action (Butler, 2004a) or via 'slippage' or ambivalence – performances that through error are done otherwise than reiterating the expected way of being that body in that place (Butler, 1997).

Butler's theories have resonance with a range of operations of power – gender, sexuality, race/ethnicity, sexuality and dis/ability. Bourdieu, helpfully, reminds us of the enduring importance of differential access to capitals and the workings of the capitalist economy in producing inequalities. Using both authors in conjunction highlights the intersectionality of these 'axes of difference' – for instance, as a white, formerly working-class and now middle-class, non-disabled, heterosexual, highly educated woman, it would be impossible to highlight which of these aspects of my identity matter most. Bourdieu also reminds us that these other 'operations of power' can have effects in terms of material inequality *and* that differentials tied to social class are enduring and pervasive. Both authors have their limitations;

however, working with my interpretation of both provides a generative context for understanding the tensions between dynamism and fixity: class, economics and other axes of power, material and social bodies. I contend that Judith Butler, despite her heterogeneous influences, is broadly Foucauldian in her approach.

An important context to this book is the specific subjections tied to normalising concepts of mental, bodily and socio-emotional 'ability'. These pervade ableist school spaces, impacting upon all young people as their mind-bodies-emotions are forged within and compared to expected norms of development tied to whatever curricula the exegesis of that moment in history, in space, demands. Of course, these usually and particularly impact upon young people who fall below expectations of development; these expectations are treated as universal, despite being socially, spatially, historically and politically constituted (Rose, 1990; Gallacher, 2017).

3.3.3.1 (Ab)normalisation and ability/disability

Bourdieu has helped to highlight the myth of meritocracy and the ways in which what is constituted as educational success is culturally specific and defined by certain racial and class perspectives – notably middle-class white people. In neoliberal education systems and society, ability is valorised and reified and schools are devised around a pervasive idea of a 'normally' developing child. The primary organisation of young people's formal curricular time in schools is usually into age-related classes and years; this is a material manifestation of the "normally developing child" (Hill and Tisdall, 2014; James et al., 1998) which underpins much education policy and practice. This normally developing child is sedimented into Standard Assessment Tests and other testing regimes, as young people are expected to achieve similar levels of body-mind, learning-social competencies at particular ages, to such an extent that schools' performance can be measured against this development. Thus, schools are a "central institutional means of normalisation" (Olssen, 2004: 70). Indeed, Foucault's mentor Canguilhem (1973) identifies clinics and education institutions as central sites for the emergence of the idea of the normal, where statistical 'averages' substitute for 'normal' (Foucault, 2003; Philo, 2007, see also Philo, 2012). McNay (1994: 95) argues:

> In modern society, the behaviour of individuals is regulated not through overt repression but through a set of standards and values associated with normality which are set into play by a network of ostensibly beneficial and scientific forms of knowledge.

Foucault (2003) contends that normative power is "always linked to a positive technique of intervention and transformation" (p. 55), or from "a reaction of rejection, exclusion and so on" to one of "inclusion, observation, the formation of knowledges, the multiplication of effects on the basis of the accumulation of knowledge" (p. 48). Nonetheless, it is evident that falling outside of 'the normal' can also lead to processes of exclusion and exile.

The reification of ability and entrenchment of the "normally developing child" reproduces processes of normalisation and abnormalisation (Foucault, 2003), as some young people fall outside of, and more specifically below, expectations of normal development. Thus, the special education institution is devised to address the needs of those young people abnormalised within the shifting socio-spatial frameworks of normality that are embedded and sedimented within education institutions. It used to be that the special and general education system were more spatially disparate, but in the UK, as elsewhere, they have converged in space somewhat, although even at the height of inclusion (in the early 2000s) significant proportions of young people were educated in segregated special settings.[3]

The frameworks of normality are not fixed and given but are reproduced and reworked anew. Processes of normalisation are not fixed, but occur via specific, everyday practices within school spaces (Hansen and Philo, 2007); these processes are:

> precarious accomplishments, eked out of a myriad of uncertain practices . . . enacted through countless small decisions, on-the-spot judgements, some (but by no means all) of which coalesce into temporary [formalised, sedimented, legislated] assemblages.
>
> (Philo, 2007: 90–91)

Processes of normalisation and what is seen as 'normal' and 'abnormal' change through time and space. An oft-cited example is views towards Lesbian, Gay, Bisexual, Transgender, Queer Plus (LGBTQ+) communities which have become more expansive in the UK and other minority world nations. The broadened norms of sex and gender that we have in the UK context are unrecognisable to earlier generations or many contexts across the globe where homosexuality remains an illegal offence. This is not to deny, however, the enduring experiences of hate crime and discrimination faced by LGBTQ+ people today. The continued existence of conversion therapy in the US and in the UK (where then Prime Minister Theresa May vowed to ban the practice in 2018, although it still remains legal in March 2023) demonstrates just how deep-rooted homophobia and heteronormativity continue to be, even within the UK, not to mention the continued repression and illegality of diverse sexualities throughout the globalised world.

3.3.4 Young people's powerful socialities: the power to challenge and change enduring inequalities

Given that performances are always provisional and a moment of improvisation, they always contain the potential to do things in other ways. Norms shift through *slippage* and through performances which challenge 'the norm', along with concerted and orchestrated political endeavours. Shildrick (2005) also emphasises the role of 'radical alterity' in being proud to have a different bodily (or mental) morphology and expanding the scope of the norm. The fact that any norm is a "precarious accomplishment" (Philo, 2007) means that there is the potential to challenge and transform what is viewed as 'normal' and what is viewed as 'abnormal'. Butler (1997) also emphasises the potential to exceed exegesis of power. Here I want to reflect upon the potentials of the need for recognition to open up potentials for the transformation of relationships to difference through this process of recognition. Jessica Benjamin regards psychic relations to others as being in a constant tension between competing desires for mutual recognition and conceiving the other as outside and distinctive to the self. Psychoanalytical geographies can begin to take forward Butler's theories in ways that consider how these affect actual social-spatial relationships and the ways in which material spaces and spatialities are configured, moving beyond the representational examples which constrain Butler's own analysis of personhood and material spaces. There is always a potential to be otherwise inherent in the bringing together of others, and it is to this potential that I turn in the next chapter as I outline my concept of immersive geographies.

Notes

1 Eton is an exclusive, archetypal, boys-only fee-paying boarding school, which has educated many of the UK and global elite and powerful. Ofsted is the UK state school's inspectorate. Inadequate is the lowest inspection rating.
2 First name used to differentiate from Ruth Lupton, cited elsewhere in the book.
3 That is, 37.9% of young people with the highest level 'statements' of SEND attended special schools in 2007 (Department for Education, 2007, 2015).

4 Immersive geographies

This chapter sets out the idea of immersive geographies. The concept of immersive geographies endeavours to capture both immersion and a sense of depth and immersive geographies as more open and connected to radical new connections and ways of being. In conceptualising immersive geographies and taking the idea of immersion as the starting concept, the *Oxford English Dictionary* definition of immersion is a good introduction: "dipping or plunging into water or other liquid, and transferred into other things", or the *Cambridge Dictionary* definition: "becoming completely involved in something". It is this sense of total involvement and bodily and mental immersion in space and time that opens up spaces of transformation, the potential to collectively become something else, to be changed via the connections, which underpins immersive geographies. A definition of immersion quickly moves on to immersion forms of learning, particularly of language, whereby people totally immerse themselves in a language and culture (and often place) to learn a language (or skill) through praxis and habituation. Although I want to capture this sense of being totally submerged within something, and the chance to be transferred into other things which is inherent within immersion, this also seems to suggest an insular inward-looking focus. Therefore, whilst retaining the sense of total involvement suggested by the language of immersion, I prefer to focus on the immersive, as more active and open to possibilities.

Schools are sites of immersive learning, as they are institutional spaces dedicated to educating young people in formal and informal ways, which intersect. In this chapter, I reflect upon what is specific about schools and other spaces in which people, and in this case, young people, come together, not just occasionally and fleetingly but repeatedly and enduringly. Time is key here, along with space, and the space/time/space/time/space/time dialectic of repeated encounters in spaces which are ostensibly the same and yet performed slightly differently every time. The bringing together of young people in specific spaces repeatedly allows them to forge deep and affective connections which transform them in some way. The idea of immersive learning is often applied to generating new worlds via virtual reality technologies. This sense of open possibilities, and the ability to generate alternative worlds, is critical to immersive geographies. In schools (young) people converge to do similar things day after day, and yet every time they come together the connection and the practice are a performance; it is provisional.

DOI: 10.4324/9781003028161-4

The repeated coming together and sharing of space opens up potential for co-emergence of subjectivities and shared histories and new possibilities of being. I draw upon Liz Bondi's notion of empathy and identification and Judith Butler's reworking of Jessica Benjamin's recognition and intersubjectivity to consider possibilities for forms of relating to others that forge new types of emotional intersubjectivity beyond the enduring frames of reference of (dis)ability, class, gender/sex, sexuality, race and ethnicity, religion and so on. This has resonance to immersive learning, where new worlds are generated through virtual reality technologies, and emphasises that the real is constantly being remade. Given school spaces are simultaneously local and global, new, more empowering ways of being and connecting can have resonance beyond that particular moment in space and time, via countertopographies (Katz, 2004). As part of an embodied habitus of young people, these moments can become part of trajectories and their future connections, relationships and space/times that they move through and as they recreate the world.

4.1 From brief, surface encounters to deep, embodied, immersive, connections

Immersive geographies take inspiration from geographies of encounter. Taking as a starting point Doreen Massey (e.g., 2005) and others' insistence that space, as well as time, is dynamic, a field of study has developed to examine how the coming together of people who are 'different' in some way, or as Massey might express it, the coming together of heterogeneities, can break down barriers and challenge enduring negative representations. Much of the focus of study has been about race (although see Dear et al., 1997, for an account of disability). Most of the scholarship in geographies of encounter has focused on urban spaces and the rise of what Valentine (2008) refers to as a 'cosmopolitan turn', which "celebrates the potential for the forging of new hybrid cultures and ways of living together with difference" (p. 324), in what Laurier and Philo (2006) refer to as the "convivial city". The coming together of 'different' people and things presents opportunities to forge new connections; this potential has ignited the imaginations and scholarship of many geographers and social scientists (see Valentine, 2008 and Wilson, 2013 for an overview; also, Amin, 2006; Laurier and Philo, 2006; Staeheli, 2003). Many of these accounts take as a launching point Massey's (2005: 181) idea of "thrown togetherness", theorising how different people (and things) come together in particular 'local' spaces at specific moments in time, to forge distinctive connections across space and time at a host of interconnecting scales, from the local (or the body) to the global.

Whilst much of the focus on encounters has been on urban and 'public spaces', Valentine and Wilson focus on family and school encounters between parents, respectively. However, encounters between young people themselves and the specific opportunities and challenges offered by the 'thrown togetherness' of school spaces are relatively underdeveloped in the literature (Mills and Waite, 2018, provide an example of examining socio-spatial relations of *young people* in the context of National Citizen Service). This is despite the fact that in one of the foundational interventions in establishing geographies of encounter, Amin (2002) points to the importance of "micropublics of everyday social contact and encounter" and specifically references "micropublics such as the workplace, schools, colleges, youth centres, sports clubs, and other spaces of association" (p. 969). Amin therefore draws attention towards young people and identifies schools as key sites of micro-publics, specific moments where "cultural destabilisation and transformation" can occur (see also Hemming, 2011).

This focus on schools as key micro-publics is developed in this book, where the socialities of young people are examined for the ways in which difference is (re)produced and performed, and connections are formed that transform axes of power relations. Overall, studies of encounter have been largely celebratory, focusing upon how encounters can challenge and transform enduring inequalities and stereotypes (see also Holloway et al., 2019 for a critique). It is this positive, and perhaps overly celebratory and optimistic, potential of encounters that I seek inspiration from, whilst at the same time holding in critical tension that, whilst encounters might have an imminent potential to generate new ways of being, there is, overall, a tendency towards endurance of the differences, power relations and hierarchies between individuals and social groups along all too familiar grounds, as Valentine (2008) emphasises. Indeed, encounters are not necessarily positive, but can reaffirm negative stereotypes, and people can abject, exclude and self-segregate from 'others'. For instance, Holland et al. (2007) have found that strangers tend to self-segregate from people they do not know in co-present spaces. Indeed, Massey (2005) points out that in "thrown together" situations there is always an inherent risk of conflict. Similarly, Valentine (2008) argues that even positive encounters can be superficial and do little to transform enduring negative stereotypes or socio-spatial relationships. She argues that a paradoxical gap is opened up between practices and values. Drawing upon qualitative research in three communities, she points to enduring negative stereotyping and attitudes in spaces in which diverse groups connect. Importantly, Valentine (ibid) highlights the importance of broader socio-spatial conditions, such as the relative affluence and security or precarity of the groups who are encountering each other; she highlights that "encounters never take place in a space free from history, material conditions, and power" (Valentine, 2008: 333; see also McKittrick, 2011). This central and critical insight is taken forward in my envisaging of immersive geographies.

Valentine emphasises that being polite or convivial in public spaces does not necessarily challenge deeply held prejudices that might be shared in 'private' spaces of home. These polite convivial encounters demonstrate tolerance in public spaces; however, Valentine (2008) emphasises that tolerance expresses power; powerful groups 'tolerate' the less powerful rather than being changed or challenged by the encounter. As Valentine (ibid) states:

> Positive encounters with individuals from minority groups do not necessarily change people's opinions about groups as a whole for the better with the same speed and permanence as negative encounters. In other words, in the

> context of negative encounters minority individuals are perceived to represent members of a wider social group, but in positive encounters minority individuals tend to be read only as individuals.
>
> (p. 332)

Immersive geographies take forward this question of endurance and change, and also surface and depth. In contrast to Helen Wilson's (2013) critique of the notion of interiority and internal psychic or mind space, I suggest that Valentine's "set of values and beliefs" are not, as Wilson claims "somehow separate and formed in isolation from encounters, rendering them fixed, stable and clearly defined" (p. 460). Rather for me, Valentine helpfully points out that there is a friction to the dynamism and transformation of 'interior' thoughts and feelings.

In non-representational theory, and in geographies of encounter more specifically, there is a danger of overly representing the dynamism of interconnected and porous bodies in space, the surface perhaps, and underplaying enduring and entrenched differences, an interior world and a tendency to endurance (Neil Smith, 2005; Tolia-Kelly, 2006). There are, however, some compelling accounts which work generatively with non-representational theories to explore the enduring nature of inequalities through the dynamism of porous bodies connecting in space (Bondi, 2005; Saldanha, 2010; Colls, 2012; Pile, 2010; see also Anderson and Harrison, 2010). Immersive geographies give space for a reflective presentation of self which conceals the interior workings of the mind – which itself is not pre-social and can be challenged and transformed, and yet is internal and can be concealed. People *can* conceal the inner workings of their mind-bodies-emotions and it seems problematic to argue that there is no interior self, albeit the interior self is not perhaps unbounded by socialisation. Immersive geographies are interested in the depth of the interior workings of minds-bodies-emotions and how these are porous, connected, dynamic and are also internal and reflected as interior spaces, perhaps a sedimented materially embodied and psychic history of all our past encounters; psychoanalytical geographies can help us here.

4.2 The 'depth' of the inner self and making new connections through socio-psychic processes of recognition and empathy: deconstructing the autonomous individual

Immersive geographies are attentive to psychoanalytical geographies and philosophies as a way of investigating the inner self, which is, however, porous and connected and never experienced in isolation, inspired by Pile (2010). Pile argued that psychoanalytical geography can provide a useful resource to theorise an interior self, a self which is connected to, and forged within, specific socio-spatial contexts, which is embodied, affective and generative, with the exception, perhaps, of the deepest unconscious (Callard, 2003; cf. McIntyre and Nast, 2011; Bondi, 2014b). In a similar vein to Judith Butler's work on the psychic life of power, discussed in the previous chapter, psychoanalytic geographies have emphasised that socio-spatial processes are tied to the operations of the psyche, and vice versa.

Psychoanalytic geographies provide a scope to analyse an interior mind, but one that is embodied, porous and connected to broader socio-spatial processes.

Much early work about psychoanalytical geographies focused on the role of object-relations theory to identifying 'the same' and 'the other' which operates at a variety of interconnected scales from the individual to the national and so on. Key to the differentiation of the self from 'others' are the processes of identification and othering or abjection. Abjection is a process of rejection and disgust in which the object, or the other person in intersubjective relations, is excluded, stigmatised and expelled. Abjection is a complex psychanalytic process, which is commonplace, but through which elements of 'our' human frailties, vulnerabilities or the mucky leaky smelly and less pleasant elements of human embodiment are associated with 'other' things, and importantly, people and groups, that can then be disassociated from the self (Kristeva, 1982; Tyler, 2009). These are clearly critical socio-psychic relations which frame isolationist tendencies from nationalism to some of the stark and sickening debates about migration which have surfaced internationally as part of populist movements. Examinations of processes of defining a 'Self' from 'Other' around disability and impairment have been examined by Dear et al. (1997), in an account which mirrors Shakespeare's (1994) and Morris' (1991) scholarship about the abjection of disabled people. In line with Kitchin (1998) and Shakespeare (1994), Dear et al. (1997) explain boundary drawing and maintenance (at an individual and wider spatial level) as emerging from an ontological need to protect the 'Self' from the 'Other'. They claim that this boundary drawing is not purely an individual/psychological event but is socially contextualised in wider society. The following quote epitomises and captures processes of abjection of disabled people:

> What is happening is that non-disabled people are getting rid of their fear about their mortality, their fear about the loss of labour power and other elements in narcissism. The point I am making is that disabled people are the dustbin for that disavowal.
>
> (Hevey, 1991, p. 34), cited by Shakespeare (1994:298)

It is intriguing how these socio-spatial processes of 'same' and 'other, or abjection and/or misrecognition (Fraser, 1998) operate to forge material socio-spatialities from intergroup connections to walls around nation states. Dear et al. (1997) emphasise that social groups attempt to build (spatial) boundaries to protect themselves from 'deviant others', and these boundaries prevent spatial proximity and acceptance of the 'other' group.

Geographies of encounter and social policies that endeavour to bring together 'different' groups depend upon the idea that proximity, breaking down the material boundaries that separate different people, will produce new connections between heterogeneity. Dear et al, and others, argue:

> Physical proximity weakens the bases of distancing as it forces a direct confrontation with disability, challenging the stereotypical anxieties that structure the diametric opposition between the abled and the disabled.
>
> (Dear et al., 1997: 474)

Dear et al. specifically relate their arguments to disability, and this has been pivotal to my work; however, these arguments have a broader resonance with any and all intersecting axes of power relations.

Yet, the question remains whether a simple putting together of different people will challenge enduring socio-spatial-psychic boundaries. The whole point of reflecting on psychoanalytical geographies is, arguably, to reflect upon the interiority of the mind. This is a mind, however, which is tied to embodied habitus. It is dynamic but not instantaneously dynamic, and its interiority or habituation has a resistance to change. In particular, the concept of an unconscious gives the sense of an interior within the interior of the mind. Butler (1997) positions the unconscious as "a certain unharnessed and unsocialised remainder . . . which contests the appearance of the law-abiding citizen" (p. 88). Butler (1997) further contends that the unconscious evades normalisation. Pile (2010) similarly discusses the idea of the 'unconscious' as an interior and deep topography of the mind, as being somewhat unknowable, which:

> [c]arries out a kind of guerrilla warfare with those agencies (such as the Super-Ego) that try to prevent it from gaining expression. The unconscious struggles to find ways of making its presence felt against all means of preventing it from so doing.
>
> (p. 14)

The idea of the unconscious, or the dark recesses of the mind, is useful to conceptualise the way in which deep psychic processes can have an influence on social life (see for instance McIntyre and Nast, 2011). Embodied reactions to others can be mediated by assumptions which defy our conscious attempts to be otherwise; reactions to others can be mediated by ways of relating to the other which can seem natural and yet more likely have been habituated through previous encounters, which defy simple transformation. However, for me, whilst the unconscious might seem to defy transformation, it is not perhaps beyond socialisation but has emerged in relation to specific social and spatial contexts. The point of reflecting on the subconscious is to emphasise that whilst people become and are transformed via connections to human and non-human others (which have their own histories, networks, mobilities and connections), we are not immediately or instantaneously dynamic. There is an interior, and to that moment, anterior, self, which is reflective, a sense of self and which is not instantaneously dynamic and shifting. There is a sedimented material, embodied and psychic personal history and trajectory, a habitus perhaps, which is present beyond, before and ahead of the immediate moment of connection and which is dynamic but not instantly or, perhaps, infinitely, so.

Psychoanalytical geographies which focus on same and other, or identification and abjection arguably take as a given an individual self which is capable of conceiving of, or living independently of, the other. This is arguably inherently masculinist and does little to deconstruct a sovereign, liberal agent. Interiority is, for me, important. Yet, this interiority does not need to predetermine a sovereign and individual agent. Psycho-social processes of recognition and empathy recognise the interior psyche or 'self'; nonetheless, they emphasise that this psyche emerges through relationships between others which transcend dualistic categorisation, through intersubjective and interdependent relations with others. This presents ways of relating which have the potential to forge new ways of being and relating. These are inherently more positive socio-psychic relations, which, perhaps. when repeated through regular encounters in immersive geographies, for instance in schools, can forge new shared embodied histories that challenge and transform enduring representations of 'others' with imminent political potentials to be otherwise.

There are two, arguably, interrelated concepts of recognition and empathy, which I explore in the following as providing insights into new potential ways of relating and becoming, but ways which do not deny the interior self. In her books *Undoing Gender (2004a)* and *The Psychic Life of Power (1997)*, Judith Butler is interested in the concept of recognition. As I emphasised in the previous chapter, the emotional need of (at least many) people for recognition is pivotal to the formation of a sense of a coherent self, which is actually, and from the outset, intersubjective and interdependent, porous and connected through the need to be recognised by the other and to live in emotional (and material) interdependency. This interdependence provides a context for subjection. Of relevance to this chapter, however, I want to examine how recognition also provides for the possibilities and potentials for transformation and 'exceeding' subjection through interpersonal connection.

Recognition "takes place through communication . . . in which subjects are transformed by virtue of the communicative practice in which they are engaged" (Butler, 2004a: 132). Thus, recognition involves an opening up of people to intersubjective dialogue that can transform all/both engaging subjects. Benjamin views psychic life as "vacillat[ing] between relating to the object and recognizing the outside [O]ther" (Benjamin, 1998, cited in Butler, 2004a: 133). Benjamin contends, although psychic destruction is possible, it is possible to develop an 'intersubjective space', which is beyond dualistic psychic relations between the Self and the Other, and which provides a context of radical openness which can transform socio-psychic relations. This openness can transform relations between people who are in some way 'different'.

Bondi's (2002, 2005, 2014a) illumination of empathy and identification similarly opens up the possibility of producing an intersubjective space wherein difference can be transformed. Although Bondi's discussion of empathy focuses upon its potential for conducting research, her illumination of empathy also has a role in understanding how young people playing – either actually in their games or as they 'play'[1] at the performance of their subjectivities through chat and the forging of social groups – can open up an 'intersubjective space' in which a new understanding of the other can be forged. Bondi emphasises:

> Empathy does not generate direct or perfect apprehension of the subjective experience of another. Rather it requires effort and is always imperfect and faltering. However much of the experience of the other is accurately recognised, empathy also entails acknowledging that the effort to understand can only ever yield an imperfect grasp of what the other feels.
>
> (Bondi, 2014a: 40)

Although Bondi is clear that empathy is always imperfect, since we are ontologically unable to entirely understand another's full subjectivity, empathy and recognition provide radical possibilities for transformational relationships to the other, and there are three interconnected elements of this. First, recognition and empathy provide more positive framing for ways of relating to people and things (or parts of people and things) that are recognised as distinct or different from the self. Second, empathy provides a relationship of trying to imaginatively and figuratively view the world from the perspective, thoughts and feelings of another. Such an act or series of actions of trying to be in the skin of 'an-other' cannot surely be easily pulled back from, into an act of denigration or re-inscription of hierarchies around perceived differences. Thus, we might recognise someone else as different, but that might not be viewed as otherness. Finally, Butler's reworking of the concept of recognition fundamentally destabilises the concept of a coherent self, since subjectivity emerges out of a relationality with others rather than (only or principally) a need to differentiate the self from the other. This means that connections can be made with others which are formed on a continuum, whereby the self is not necessarily understood as entirely separate from the other. Such an approach can radically destabilise *any* ontological separation between different categories of being – myself, yourself, disabled, non-disabled and so on – with a more fluid continuum. Close observation of the ways in which young children in particular perform their interconnected bodies gives further weight to the idea that young children do not see their bodies as contained and bounded. I commissioned my daughter to try to capture this interconnectedness in a drawing (see Figure 4.1).

Exactly when relationships of empathy or recognition emerge, rather than abjection and othering, is not clear and neither is it necessarily predictable, although broader socio-spatial context does matter, as discussed in Chapter 7. In addition, I argue that time and repetition in spaces that are shared with regularity are critical to forging positive psycho-social relations, as repeated performances forge new collective histories and memories that facilitate intersubjective recognition.

Figure 4.1 Children's connected bodies in the playground

Source: Amelie Smith

This sense of repeated encounters and circular time is encapsulated in immersive geographies.

Geographies of encounter have arguably struggled with space and time at the same time as foregrounding them; the fleetingness of potential encounters against an enduring tendency for embodied differences to be reaffirmed and endure through time and space presents a conundrum. It has been a challenge to address enduringness/interiority against dynamism/surface. Another challenge is to consider how the scale of the body and co-proximal encounters in small scales is tied to broad-scale inequalities that can be observed and are empirically measurable at a variety of levels, from the individual to the global. This tension is encapsulated in the critique of Valentine by Wilson (see Wilson, 2014) and Valentine's search to conceptualise an interior psyche which is resistant to immediate transformation. This tension is also present in critiques of non-representational theory and theories of affect by post-structuralist feminists and others. In the following, I begin to reflect upon a potential way forward in this tension through immersive geographies.

4.3 Immersive spaces

Schools provide specifically interesting spaces of immersion, as a site of depth or total covering, wherein encounters are repeated day after day; spaces in which "encounters accumulate, to gradually shift relations and behaviour over time" (Wilson, 2017: 463). Whilst embodied subjectivities are dynamic, the dynamism is within hierarchical frames of power which more often reproduce the same inequalities and power differentials rather than forge new ways of being. Within this context, everyday performances within small-scale spaces such as classrooms, playgrounds and lunchrooms are connected to broader cultural, social, economic and political processes in ways which frame, constitute, constrain and enable these performances. Importantly, these broader contexts also forge the bodies and subjectivities of the people in these places. These embodied subjectivities are dynamic and change in response to the ways they come together with others (human and non-human) in specific spaces and particular times. They are not, however, infinitely dynamic, and they are differently positioned in relation to where they are and where they have been in space/time and their myriad connections to a host of processes (interconnected representations, economic, political, social, material processes) operating at a range of interconnected scales.

The specific nature of the space of schools (or perhaps any space where the same people come together repeatedly through time) matters to immersive geographies. Gregson and Rose (2000), Massey (2005), Allen (2011) and Katz (2018) amongst others (e.g., May and Thrift, 2001) have been critical to emphasising the intersections of space/time with understandings of how power operates in both productive and more disciplinary ways (Philo, 2012; Doel, 1996, 1999). Holloway and Valentine (2000b) made a landmark intervention by drawing upon this nascent literature to emphasise the contribution that spatially sensitive, geographical, accounts can make to social studies of children and young people. This intervention was important in ensuring that social studies of children and young people was, from the outset, interdisciplinary and had a spatial bent. More broadly, the profound influence of these and other scholars (e.g., John Urry) has instigated a 'spatial turn' in the social sciences (Sheller, 2017) and sociologies of education specifically (Lingard and Thomson, 2017), with a heightened sensitivity to the power of space/time, place, scale and, more recently, mobilities. There are five interconnected aspects of the spatiality of schools which are pertinent to immersive geographies: schools as institutional spaces; schools as heterogenous sites of diverse 'cultures'; schools as spaces of power; schools as provisional, performative, becoming, spaces; and, finally schools as connected, open, porous spaces which are specific moments, interconnected at a variety of scales from the body to the global to circulations of powers, resources, ideas, materialities and things. In the following, I briefly reflect upon these aspects of the spatiality of schools.

First, schools are institutional spaces for the containment of young people, with specific and relatively codified rules and regular time/space activities. Young people are expected to attend schools (or other forms of education) and the day is structured to provide timetabled activities dedicated to various activities tied to what the exegesis of the day considers it to be essential for children to learn. Curricula are powerful, political acts, rather than being natural, and these have real impacts on the time-space of young people, families and educators' lives (see for instance Klein, 1991; Tomlinson, 2005). A critical moment in the UK was the development of a National Curriculum in 1989. This set out the subjects that students would learn, and increasingly the way subjects are taught. This and similar initiatives such as the literacy and numeracy hour in the UK reflect political interventions that have direct consequences for young people's time/space experiences, and also contain powerful messages about the skills, attributes and abilities that are important for young people to acquire, setting up new divisions of success and failure around these normative values. For the most part young people are grouped according to their age, thereby instituting a central figure of a "normally developing young person" who is expected to reach particular levels of cognition and development at specific ages (James et al., 1998). This disables any young people who fall outside of, and typically, below normative age-related expectations of learning. What starts out as a statistical average becomes a regulatory norm. Ease of administrative management and organisation has become a powerful categorisation tool for young people, which, in being redressed via 'Special Educational' programmes, also becomes resource-intensive, stigmatising for young people, and ultimately is failing young people who fall outside of age-related milestones (Azpitarte and Holt, 2023). Critical for immersive geographies is the regularity of the coming together of the same group of young people and adults to do the same broad subject or activity repeatedly over time. It is also telling that material spatialities and temporalities frame, constrain and influence young people's encounters and help to forge particular subjectivities, such as more or less 'able' (see Chapters 3 and 8).

Second, schools are heterogenous sites of diverse 'cultures' – cultures of formal and hidden education and young people's various cultures, which are active, recreated through everyday practices and in relation to broader social and cultural processes. Schools have formal cultures of learning and instruction in which young people learn the subjects considered to be critical for their futures. In addition, and simultaneously, schools are sites of the coming together of young people's cultures, in which they share their ways of being in the world, which are intimately connected to broader 'cultures', derived from families, other friends, the media in all its forms, popular cultures and so on. Finally, schools are sites of hidden curricula, which crosscut and connect formal and informal cultures. These hidden curricula teach young people how to be appropriate subjects or citizens in both deliberate and conscious (e.g., citizenship education, Pykett, 2007; or emotional and social education, Gagen, 2015) and in implicit and beyond-conscious ways. Much of the book is concerned with this element of the spatiality of schools.

Third, and connectedly, schools are sites of power. There are various ways in which power operates in schools, and power is diverse. Clearly, schools are panoptic spaces of the surveillance and control of young people, in which young people are expected to increasingly self-regulate their bodies/minds to become appropriate performing subjects. Vision is often prioritised in the disciplinary power of the panopticon, as the authoritative adult can sweep their gaze around the classroom, casting their glance upon a child at any moment, and hence the children behave appropriately (or keep their resistive practices subtle). There are other ways that regulation and control work within schools, with Gallagher (2011) highlighting the role of sound in the control and regulation of young people. The marking of books and surveillance of work is another way in which young people's time and space, even beyond the classroom, is surveyed. Schools are sites of normalisation, as discussed in Chapter 3. Schools are not discrete entities in this operation of regulatory power; rather, they are connected to a broader education institution which regulates the activities within schools through regimes of testing and inspection, for instance. In schools, power is played out creatively in forging young people into subjects that are recognisable and deemed as permissible. These can be formalised processes of subjectification or subjection, such as via citizenship education or social and emotional education. These can also be beyond-conscious positionings of young people and adults to each other and themselves. These ideas are central to immersive geographies and are developed further in the subsequent chapters.

Fourth, although sites of regulation and normative operations of power, schools are also dynamic and performative spaces of opportunity and openness, or as Philo and Parr (2000) expressed: "precarious" (p. 518) "geographical accomplishments" (p. 517), which I have morphed into precarious geographical accomplishments. Power is never absolute or complete, and as people and things come together in specific moments in time/space, despite the regularity, there is always an opportunity. Every encounter is a performance, and recreates the space/times again, and there is, therefore, always the possibility to do things slightly differently. Immersive education emphasises how new 'virtual' worlds can be generated via new immersive virtual reality technologies (De Freitas et al., 2010). Taking as a launching point that idea, I want to consider how, if every coming together in time is provisional, underpinned by uncertain and precarious performances, young people are creating new worlds every day. Herein lies the potential of immersive geographies, a potential for change and transformation, or doing things differently. Indeed, this follows the lead of Massey (2005: 11), who emphasises that:

> space is always under construction. Precisely because space on this reading is a produce of relations-between, relations which are necessarily embedded material practices, which have to be carried out, it is always in the process of being made. It is never finished; never closed.

Similarly, as Gregson and Rose (2000: 441) assert:

> We want to argue that it is not only social actors that are produced by power, but the spaces in which they perform . . . performances do not take place in already existing locations: the City, the bank, the franchise restaurant, the straight street, [the school]. These "stages" do not preexist their performances, waiting in some sense to be mapped out by performances; rather, specific performances bring these spaces into being. And, since these performances are themselves articulations of power, of particular subject positions, then we maintain that we need to think of spaces too as performative of power relations.

Importantly, this sense of openness and creativity does not jettison order. This is "no collapse into total indeterminacy; rather it is what Sadler (1999) expresses as 'a more multifarious order'" (Massey, 2005: 117).

Fifth, schools are specific institutional spaces, with connections to a particular set of institutional practices, processes, rules, laws even and expectations, and with relatively codified expectations on behaviour and time-space regularity. Nonetheless, they are simultaneously, as 'geographical accomplishments', specific moments in time/space, specific points on a map, particular places. As such they are both unique, having meaning imbued in them, and, yet, the 'sense of place' tied to a specific school is generated both by its specific history and trajectory and also by its porosity and connections to materialities, ideas and flows at a variety of interconnected scales that move through and come together in the "thrown togetherness" (Massey, 2005) of the school. Schools as spaces are "a simultaneity of stories so far" (Massey, ibid: xi).

All of these interconnected aspects of the spatiality of schools are important to immersive geographies. Indeed, arguably it is the very disjuncture and rupture between the regularity and predicable nature of the space/time of schools and the fluidity and provisionality and potential dynamism of these performative spaces which provides both the tendency to the same and the potential to transform which is inherent to geographies of encounter. The fact that schools are porous and connected spaces point to wider potentials for transformation that can exceed the specific space/time of the school, as discussed in the next section. First, however, I want to turn to the question of time. Time and repetition are central to immersive geographies.

4.4 Immersive time: space/time/space/time/space/time repeated

Time is critical, and of course iterative with space – variously understood as space/time (Massey, 200) or time-space (May and Thrift, 2001). Nonetheless, geographies of immersive geographies stand as a corrective to geographies of encounter which are often about spatial proximity but have less to say about repetition, circularity or the 'depth' of a shared history. The circular repetition of the daily, regular encounters of young people in schools opens up potential for new histories and memories which can generate affirmative ways of being, and a shared co-created history and knowledge of each other can, perhaps, promote recognition and, potentially, empathy. The regularity of the repetitions of the same people being in the same places doing similar things repeatedly, performing their subjectivities collectively, learning more and more about each other's subjectivities through the passage of circular time as similar events repeat themselves over and again, opens possibilities for recognition and empathy which would arguably not be possible within fleeting contacts in public spaces. As an improvisation, there is always the chance for doing things differently (either deliberately or by accident) and thereby challenging and changing enduring embodied inequalities.

This repetition and regular meeting provide a context to forge shared social and collective histories and new embodied subjectivities, providing the possibility of forging connections that transform difference and/or performing differences in more affirmative ways. Further, these new ways of being could generate connections which challenge, question and destabilise the reification of such enduring 'labels' as disabled. The work of Cockayne et al. (2020) intrigues me as it deconstructs taken-for-granted axes of power relations or subjectivity and challenges us to find connections across myriad differences. The categorisations of difference are often left unchallenged in accounts of encounter. Cockayne et al. (ibid.) draw upon Deleuze's "Möbius topology" to categorise space 'as difference'. Deleuze's ontological questioning of difference holds an important insight into the potentials to be otherwise, and Cockayne et al. point to some important potentials of the imminence of the encounter not only to forge new connections across difference but to radically destabilise the enduring sedimentations of particular axes of difference, forging connections which deconstruct and challenge differences, or that are beyond the self-same axes of power. They argue:

> Möbius topology . . . that invites geographers to approach the encounter as simultaneously a geohistorical production and an immanent spatial event through which difference continually emerges.
>
> (p. 194)

By focusing on skin and on the topographical view of space, Cockayne et al. (2020) helpfully provide tools for thinking through the potentials for transformation via encounters. The focus on the skin and touch as a connector provides important insights into ways to forge new connections which can transcend both difference and liberal senses of a rational, solitary, independent subject, and these could be argued to be forms of 'inter-embodiment' (Lupton, 2013). Nonetheless, the focus on the skin is also to reaffirm the surface and deny interiority (thought, memory, the embodiment of previous encounters). This reproduces what I consider to be a central dualism in the existent literature, between that which focuses upon a sense of endurance (of difference) of stark inequalities (in line with Valentine) and those accounts which overemphasise dynamism, co-presence and the potential for change, with Cockayne et al. (2020) falling on the latter side of the divide.

Nonetheless, Cockayne et al (2020) conceptualises both a circularity and linearity of time, which is interesting and productive. Without using the complex language of Deleuze, I am rather going to simplify by saying that in "Möbius topology", circular time and linear time come together in an infinite and dynamic relation in particular spatial moments. For me, it highlights two important and interconnected elements of space/time and immersive geographies: the circularity of time (or rather space/time) and the linearity of time. The circularity is about repeated performances of the same people (and things) coming together ostensibly performing the same things repeatedly; yet every time is an improvisation and can differ in subtle or greater ways. This repetition forges familiarity and can facilitate connection, empathy and shared subjective histories. The linearity expresses time as a line to the future, comprising myriad circular moments, but also in which these repeated moments forge a particular path or trajectory, as, for instance, the young people age and time moves on. Much like the mobility of the vibrating molecules that coalesce to give the appearance of cohesion of a body, these repeated, circular, motions through time are experienced as a trajectory or history when reviewing in retrospect. As Massey (2005) emphasises, on the "Codex map[2] – the directionality of the footsteps makes it clear that there is no reversibility here: you can't go back in space-time" (p. 109).

These moments in space/time only emerge because of the specific circularity of time, drawing upon performativity theory (Butler, 1990), which sees subjectivities, and spaces (see Gregson and Rose, 1990 discussed earlier), emerging from repeated performances over time. The repetition over time and space is of interest: the repeated, daily encounters of things being the same yet different – a specific moment in space and time, and yet also the same people coming together to do something ostensibly similar, scripted and timetabled, time and again. Connections, which are at once about connecting porous bodies through touch, play and co-presence, but which are also about depth, about learning to understand the interiorised thoughts, feelings, sedimented history of the mind-body-emotions of young people. Repeated connections, which facilitate intersubjective relationships and the forging of new shared histories, memories and connections. Connected performances, which appear to do the same thing, but which every time are an improvisation, are emergent and can make the world anew as is expressed in immersive geographies.

In the space/time/space/time/space/time of the school, the potential is in how in topographical space (Cockayne et al., 2020), things connect. This can allow the forging of new "bodies without organs" (Nosworthy, 2014). These can deconstruct, challenge and work anew categories of 'the same' and 'the other', by "blow(ing) apart strata, cut roots, and mak(ing) connections . . . rhizome-root assemblages, with variable coefficients of deterritorialization" (Deleuze and Guattari, 1988: 17, cited in Holt and Philo, 2023). An example here is the body without organs, or the assemblage of four children and a wheelchair in which the child in the wheelchair is connected to four other children, leading them in a game of horse and carriage to emergent possibilities, where a wheelchair is an object of potentials and possibilities rather than a symbol of limitation and tragedy (see Chapter 7). These bodies coming together are not just 'surface bodies'; they are *contextual bodies/subjectivities/agencies*, with an inner and interior life, which has emerged from their genes and materialities in dialogue with past encounters in past space/times and become anew through new inter-embodiments with people and things, which have their own history, and future trajectories.

Young people as *contextual bodies/subjectivities/agencies* are becoming within specific socio-spatial contexts and are already the embodied memory of their origins, albeit they are still dynamic and open to change and new possibilities. Their coming together in schools provides new connections and the potential to forge affirmative collective subjectivities or habitus, and every coming together provides imminent potential to be otherwise. Every repetition is an improvisation and provides alternative ways of connecting which provide ways of transforming or performing subjectivities in new and creative ways; however, these do not happen in a vacuum but occur within specific moments in space/time. Further, these new ways of being might have a potential beyond the immediate space/time of its performance, in the very least as these young people take their embodied subjectivities elsewhere in future time/spaces. There is something intriguing in the intersection between the individual experience and trajectory and how it connects to and might represent collective experience and trajectory. Immersive geographies can seem very local and specific, yet at the same time, these moments in space and time are porous and connect to other space/times. Therefore, these creative ways of being or new worlds that are forged might have resonance beyond the specific school space at the same time as being framed and constrained by them.

4.5 Immersive geographies generating new worlds beyond the school

At the same time as the performances of young people within schools are specific, repeated moments, in particular spaces/places/times, these spaces/places are also unequally connected to a range of socio-spatial processes emerging from a host of scales from the local to the global. Schools are places and can be seen as unique points on a map (two schools or even two classrooms are not exactly the same, although they have similarities). Nonetheless, these places are specific not (only) because of their inherent bounded characteristics but due to their relationships between things and people within them, their porosity and their heterogeneous connections beyond that specific place. As Massey (1993: 67) sets out, a view of earth from a satellite would show, if it were possible to visualise them:

> Economic, political and cultural social relations, each full of power and with internal structures of domination and subordination, stretched out over the planet at every different level, from the household to the local area to the international. It is from that perspective that it is possible to envisage an alternative interpretation of place. In this interpretation, what gives a place its specificity is not some long internalized history but the fact that it is constructed out of a particular constellation of relations, articulated together at a particular locus. If one moves in from the satellite towards the globe, holding all those networks of social relations and movements and communications in one's head, then each place can be seen as a particular, unique point of their intersection. The uniqueness of a place, or a locality, in other words, is constructed out of particular interactions and mutual articulations of social relations, social processes, experiences and understandings, in a situation of co-presence, but where a large proportion of those relations, experiences and understandings are actually constructed on a far larger scale than what we happen to define for that moment as the place itself, whether that be a street, a region or even a continent.

Massey's later work puts more theoretical flesh on the bones of this pivotal and transformative insight. Of course, the nature of space and time has continued to be transformed since 1993 and even 2005. Most notable here is how the distinction between 'virtual' and 'real' co-presence would seem to be no longer viable, given most of the globalised world is typified by having simultaneous virtual and real hybrid co-present experiences, for example, hybrid work meetings, glancing at social media on the phone, whilst streaming on the TV, posting on social media or responding to emails whilst on the beach with family and friends. Communication technologies, intensified by the Covid-19 pandemic, have led to a "time-space compression" (Harvey, 1989) in which people can have virtual co-presence in real time in an experience which is embodied and has most elements of physical co-presence, although as yet we cannot touch or smell or have the full affective experience of co-presence through Microsoft Teams or Zoom. The research in this book captured some of this transformation (although not the intensification post-2020), and many of the young people in the study connected with a host of others via technologies such as social media and gaming consuls.

Topographical concepts of space and place emphasise the interconnections and power relations that defy fixed notions of scale that hierarchically move up from the body to the global. Rather, topographical conceptions of space utilise 'flat ontologies'. Similarly, these ideas of space/time highlight the duality of local/global the global directly impinges on the local, or even bodies, and that the 'global' is inherently 'local' in origin (Massey, 2005; Allen, 2011; Ansell, 2009). In these conceptions, then, socio-spatial processes can 'skip scale'. This provides imminent possibilities for the potential for things that are happening in particular places and moments to have impacts and influence beyond their points of origin; these possibilities are critical to immersive geographies. Katz's (2001, 2002, 2018) notion of countertopographies teases out these imminent potentials. Countertopographies draw 'contour lines' between the common socio-spatial processes impacting upon differently circumstanced places, to demonstrate:

> the connectedness of vastly different places . . . which reproduce themselves differently amidst . . . common political-economic and sociocultural processes.
>
> (Katz, 2001: 1213)

Katz points out that such different outcomes of common processes, provide insights into possible alternative futures:

> the notion of countertopography is meant to invoke the connections between particular historical geographies by virtue of their relationship to a specific abstract social process or relation, such as restructuring or deskilling. New political economic formations can be mobilized along "contour lines" connecting particular places (or topographies). The politics sparked and informed by countertopographies – by making unexpected connections among disparate places – can produced spatialized abstractions akin to those fostered by the permutations of globalizing capitalist production . . . . countertopographical analysis may have the fluidity to match the deft moves of capital and its attendant social relations, and thus expose their wily ability to produce uneven developments across space and scale in ways that eclipse and hide common grounds in a welter of difference and inequality.
>
> (Katz, 2004: xv)

There is a substantial and important literature that draws upon countertopographies to reflect upon ways of being and doing that defy or challenge the dominant global world order, especially global capitalism and neoliberalism(s) (see for instance the special issue edited by Pain and Cahill, 2022). They tend to engage with historical materialist or Marxist approaches, which have been somewhat sidelined within English and Northern European studies of children and youth. These accounts give important insights into enduring and intersectional, intersecting, cultural and material imbalances and disadvantages. Emphasising and critiquing the operations of global capitalism has never been more important and has often been sidelined within social and geographical accounts of childhood. Nonetheless, overthrowing the capitalist order seems ambitious. Indeed, even Harvey (2020:13) emphasises that social reproduction is so implicated and connected to global capitalism that it is neither possible nor desirable to destabilise global capitalism in one foul swoop; rather he suggests that:

> The task is to identify that which lays latent in our existing society to find a peaceful transition to a more socialist alternative. Revolution is a long process not an event.

Countertopographies also give insight into more modest social justice possibilities. From a different, although connected, standpoint, which draws upon topographical ideas of space and flat(ter) ontologies, Kitchin and Wilton (2003), provides an excellent example of how activists (in this case disability activists) can connect via new technologies to forge connections across space/time, in places with diverse histories and geographies. These new connections and shared knowledges provide insights into shared experiences of oppression and find strategies to challenge and change these oppressions.

Here, reflecting upon these conceptualisations, I want to posit the idea that if everything that is global is simultaneously local, it is possible that something that challenges and changes representations and performances of difference in a specific, small scale, point in space and time, can have resonance far beyond the specific context of its emergence. As Massey (2005: 182) points out:

> each local struggle is already a relational achievement, drawing from within and beyond the "local", and its internally multiple . . . The potential, then, is for the movement beyond the local to be rather one of extension and meeting along the lines of connected equivalences.

Is it possible then, that some more affirmative performances presented in this book provide alternative ways of being and doing that deconstruct difference as otherness or, even more ambitiously, reframe difference and connection around a continuity of mind-body-emotional types that have resonance beyond the immediate time/spaces in which these were played out?

The book can be seen as a countertopographical project, which both demonstrates how young people in different schools experience similar processes of subjection around powerful and enduring axes of power and suggests alternative ways of being. The book foregrounds moments in which young people exceed and challenge the exegesis of power, and become something else, radically decentring regimes of disablism, class power, and racial and ethnic inequalities. For the most part, these ways of being are moments in space and time unconsciously performed and were just about playing – playing something completely separate and independent from 'real' life or playing with subjectivities and ways of being in the world. The young people in the book were largely not actively activist and deliberately countering oppressions, although there are some examples where the young people are touching upon this. The immersive moments with radical potentials to be otherwise were not self-reflected upon by the young people, for the most part. They have been taken away and reflected and analysed by me and others. I have also reflected upon the conditions of the emergence of these moments in which new, emancipatory, potentials emerge to begin to reflect upon whether these can be recreated in other space/times (see Chapter 8).

Another way in which these practices are pertinent beyond the immediate space/time of that moment in that school is that as young people are becoming (Uprichard, 2008), they are frequently conceptualised as embodying the future of societies, and in a very real sense they do. Young people's bodies are inherently part of a "prefigurative politics" (Jeffrey and Dyson, 2021: 614). Future time is always present within young people's bodies and their co-presences in the institutional spaces of schools. This does involve 'improvisation' (Jeffrey and Dyson, 2021); however, it is a mundane and everyday improvisation. A grounded everyday politics which has imminent potentials to be otherwise. This may be tied to concerted political framings of an alternative view of utopian, or at least more affirmative, future social possibilities. More often, perhaps, it is just about young people playing their identities in different ways which have transformative possibilities, which might not be at all obvious to the young people doing the performing. Young people's bodies, then, are at the nexus of the future and the present. They are also a component in intergenerational relations. Enduring (dis)advantages are reproduced through young people's bodies-minds-emotions, which interconnect with intergenerational habitus emerging from home environments. As becomings, young people are not just the beings that they are in the present, they are also their future selves as they grow, and the events of the moments are embodied within their subjectivities and taken forward in some way in their adult futures. It might be that these affirmative moments become part of a habitus and an unconscious backdrop framing future social interactions. The powerful potentials of young people's social relationships in school spaces are examined in the following three chapters.

Notes

1 Aitken and Herman (2010) and Katz (2004), among others, also talk about the generative potentials of play and playing.
2 Massey (2005) discusses The Codex Xolotl, on which the reader follows footsteps to trace the trajectories of events through time, which she points out are "representations of time and space together" (p. 7).

5 Young people's friendships

Embodied, emotional and social capital

In this chapter, the experiences of young people are presented, although clearly with the caveat that these are my/our representations of the discussions, observations and, where appropriate (yet less pervasively) the creative artefacts, of young people – including my children. In this chapter, I examine how young people's friendships are critical to them for providing emotional support, social and emotional capital and 'recognition'. The chapter also witnesses how foundational young people's social relationships are to them, but also how fraught and provisional, the constant work required to make and forge friendships and the dynamism, instability and fragility of friendships, some of which endure despite this need for constant affirmation and work. The need for recognition and connection with others demonstrated in the aforementioned discussions means that most (young) people, as *contextual bodies/subjectivities/agencies* are being forged in connection with others. Therefore, young people's social relationships are powerful, foundational even to many young people. In addition to providing important emotional and social capital, this need for recognition also forges young people within frameworks of power and subjection, and how this aligns with enduring axes of power is reviewed in the next chapter.

5.1 Friends, good friends and best friends: emotional reciprocity, connection and affirmation

> you and me are very close – we're like sisters kind of and well, because we don't have sisters we're like sisters.
>
> (Mahal[1] and Jasmin[2,3])

The importance of friendships is expressed in the plasticine model made collaboratively by Mahal and Jasmin to represent their friendship. Nearly all the young people who participated in the study with the whole range of mind-body-emotional characteristics in every type of school discussed having friends. Young people emphasised the importance of friendships and good friendships in emotional and emotive terms; as Mahal and Jasmin emphasise, they are

DOI: 10.4324/9781003028161-5

Figure 5.1 The plasticine model made by Mahal and Jasmin

close, like sisters, with close emotional ties, empathy and identification, or as Paavai[4] states:

> Yeah, a friend is someone who you can trust. It's a two-way relationship, so you do things for them and they do things for you. And yeah, as Emilia said, it's someone who you can turn to when you're in need.

Erin: I think they're really important because you have someone to turn to who is not, if you don't want to talk to someone who is like in your family you can turn to a friend I guess!

Samia: Yeah, and you could tell them anything and know that they won't judge you for it. So yeah that's good.

Erin: Sometimes they judge (all laugh) but it's only in a fun way!

Saabira: Someone you can have fun with and a laugh.
(All giggle)

Cora: And someone you can trust with secrets and know that they won't go running off with . . .

Karolina: And like make you feel happy, yeah.[5]

These friendships provided important emotional and social capital, along with a sense of recognition. The social capital provided access to:

> the aggregate of the actual or potential resources which are linked to possession of a durable network of more or less institutionalized relationships

of mutual acquaintance and recognition – or in other words, to membership in a group – which provides each of its members with the backing of the collectively-owned capital, a 'credential' which entitles them to credit, in the various senses of the word.

(Bourdieu, 2018: 249–250)

The emotional capital also provided young people with a habitus of being a friend and having friends, which gave a sense of self that could be taken into other contexts (Holt et al., 2013).

All these stories of friendship are visceral in their emotional impact. The friendships of children in the middle-ages of childhood (aged 8–11) in particular are fascinating and complicated. The children who were identified as 'friends' and 'best friends' did not always map onto who we observed children playing with in the playground. For instance, Annie discusses her friendship with Rosie, yet I never observed them playing together in the playground. Sometimes friends represented an idealised notion of who children would like to spend time with, rather than their actual playmates:

Annette[6] Rosie is my best friend.
Int: Have you got any other friends?
Annette: Sara and that's it.
Int: Why is Rosie your best friend, and is Sara not quite such a good friend?
Annette: Yeah, because Rosie plays with me all the time, and Sara just plays with the boys all the time . . . . Because we don't play with Maya. She doesn't have a good laugh; she just walks around the playground.
Int: And what do you do with Rosie when you play with her?
Annette: We play funny games.

Similarly, Lucy[7] played almost exclusively with Lindsay[8] at playtimes, yet did not mention her within the list of friends she provided:

> I like playing with my friends, my friends are Emily, Claire, Jessie, Miss Holt, um – Rosie, who's at my old school, Mark's brother, another Rosie, um – Annabel, another Emily, Max, Mark, I've got loads.
>
> (Lucy, Rose Hill)

From the aforementioned excerpts, and throughout the book, and as anyone who has worked or researched closely with young people will know, friendships are complicated. Friendships can be precarious at the same time as they are often enduring. They provide crucial emotional support and connection. For most young people (and indeed, perhaps, most people) having emotional and social recognition (Butler, 2004a) is a fundamental requirement to thrive in school, and more broadly. Friendships are not incidental, they are fundamental. Friendships are always connections across difference, and in some sense forge new possibilities and horizons of being, albeit that these differences are often subtle and not always obviously demarcated to outsiders' view (see also Cockayne et al., 2020). Yet, when we question young people about who their friends are, they frequently quickly slip into who their friends are not, and most of the young people in the study had also had negative experiences of friendships falling apart, being left out and/or being bullied. Very few of the young people in the study were completely socially isolated, although a small number of young people in every type of school were, and this is discussed in the next chapter.

Those who are friends, and those who are not friends, are intimately interconnected in many young people's accounts. Friendships can define themselves in opposition to the rest. The paired interview between Emma and Laura[9] is indicative of the importance of friendships, but also how friendships are precarious, fluid and shifting:

Emma: Well Laura's, she's quite confident, not like really, really, really confident. She's very, very funny. A lot of people take her for granted and like when like people don't get to know her, because like oh she's quite quiet, so they don't get to know her, and everyone sort of thinks oh because she's like, she's OK, but you just need to try to get to know her and you realise she's really nice and funny.

Laura: Well Emma, she's really funny in class and out of class, and people sometimes think that she's not really funny and like she's just being mean when she's not. And then before, not to offend you, but I thought the same as others, but then when people come to be her friend, then you find out she's not actually trying to hurt your feelings, she's just trying to be funny, and she is.

Emma: But they sort of, but now like because it was really hard for me like because, you know Mary, well she was my best friend and then you know Marge, well just before we went to [outdoors residential centre] she like came and she wouldn't let me even speak to Mary, like she'll just like drag her away and this kind of thing, and I was just left on my own, and it was really upsetting. And then I agreed [with Mary], because our mums and dads are really friendly outside school, we've been camping together and stuff, and we were going to sit next to each other on the coach, and then Marge was just like actually me and Mary are sitting together, and Mary just followed her, and so I was kind of left alone.

Laura: And then Mary was like was, because we were, there was like loads of rooms, me, Emma, Heather and Ruth, we were all in the same room, and next door was Mary and a whole load of group . . .

Emma: Everyone else but me . . .

Laura: And they were letting, they were like in the morning, so like we were trying to sleep, they'd come and knock on our door and run off.

Emma: And then, and then they kept doing that, so we knocked on their door and then they all told of me [told the teacher].

Int: Oh, that sounds a bit mean really, not much fun. Is that quite a sort of important thing for you then, kind of going away to, is it like a residential camp thing?

Emma: Yeah, it was for two nights and three days, and it was the first time I'd been away with the school, so I was quite looking forward to it . . .

Laura: And you were upset . . . *[whispers]* she was missing her parents.

Emma: Yeah, I was really upset, so, and because she was *[unintelligible]* sit back there because I was really looking forward to like spending it with, me and Mary called ourselves best friends, but then like it all changed and stuff. So now me and Laura, so I decided to give you a turn.

Laura: Because we were both in beds next to each other, and we slept looking at each other's face!

Emma: Yeah, and Mary kind of sort of wants, doesn't really care about people, because the other day she just, she comes up to me and she just like tries to take me away from Laura and saying like come on like let's play together . . .

Laura: Yeah, it was swimming yesterday and she was like oh that swimming was really good . . .

Emma: . . . good, we're in the same group, and like she was looking at Laura with a glint in her eye because Laura couldn't swim . . .

Laura: And Emma was like well . . .

Emma . . . well she can swim but she's like . . .

The way that Laura and Emma's friendship was forged through a falling-out amongst another friendship group is intriguing. It was a chance encounter because the two girls were in beds next to each other on a school residential, when Emma needed some emotional support and Laura was there for her. This demonstrates the importance of encounters; also how critical it is to provide opportunities for social encounters between young people (see Chapter 7). The exchange demonstrates the importance of encounters and the chance of being co-located or sharing an activity or space to the opportunities to finding shared interests and/ or empathy.

The girls both have interests in common; they are both high achieving academically (especially Emma) and sporty, demonstrating a 'perfect girl' subjectivity (Ringrose, 2007; Pomerantz and Raby, 2020; McRobbie, 2009; Allan and Charles, 2014). Along with being so academically high achieving she needed to be removed from class sometimes to do extra work, Emma had written a full novel. In answer to the question of what she planned to do when she grew up, Emma stated:

> Yeah, I, yeah I don't really know what I'm going to do yet, but I really enjoy writing, like I've already written a novel and like I'm typing it up on the computer to get it published.

Intriguingly, the girls, and especially Emma, were sometimes excluded, stigmatised or bullied on the basis of their ability and success. Despite the pivotal emotional support and affirmation between the girls, in the discussion of how Mary is trying to 'pull' Emma back to her, in this paired interview we can see the centrifugal forces pulling the friendship apart. We also heard Mary[10]'s account:

> Lottie, and Lola wasn't there then. And I used to be best friends with Jade which was really weird, and then in year two I was best friends with Emma and then I, then I was best friends with Emma all the way into year four and then I didn't have a best friend and now I have a best friend again!

Even in the joint interview between Emma and Laura, we can feel the dynamism of the relationship. Friendships are maintained through everyday and mundane practices, particularly gossip and conversations and just being together, particularly for girls. The sharing and swapping of food from packed lunches was also raised as important for conviviality:

Mary: Well, I think we're both friends with everybody in the class, we're not specifically best friends with anybody, um, we've known each other for ages, since foundation, but we didn't become best friends until earlier this year.

Int: Oh ok, so how did you get to be best friends, what happened?

Mary: Well we sort of had this group with me, Eden and Jade and then um, Jade started going off and playing with Cara and Rhianna and stuff, so it was just me and Eden left and then we kept playing with each other and I just said to Eden that I wanted to be best friends with her, and she just said um, she said that she wanted to be best friends with me.

Int: Cool

Eden: Yeah, we do lots of things – we share each other's lunches, especially hers.

Mary: Well, um, sometimes we play with other people like Marge and Sally um, sometimes we just sit together and just chat . . . . [11]

It is evident that for the young people, friendships are often dynamic and shifting, provisional and yet critical. I have theorised this criticality via the concept of emotional capital (see Chapter 3). The 'capital' in emotional capital is important, as friendships provide a resource 'in the bank' that young people can take with them as a sense of validation, that they are able to forge friendships, and this has an impact by becoming an embodied, habitual confidence and a backdrop to future encounters. Emotional capital is also a reminder that friendship and the social connections that young people forge are differently positioned in relation to the prisms of capitals and the broader places in which the young people are placed (see also Chapters 3 and 7). Yet, capital does not go far enough. For most young people friendships and the emotional reciprocity, they provide are foundational to being in the world; foundational yet provisional, shifting, always open to conflict.

In the interviews and the research diaries, we only see the surface of the young people and only hint at their depth. What we see is what they present to us – performing what they want us to see, which can be an exaggerated version of how they have been typecast: clever, rebellious, sassy, sporty and so on. It must have been devastating for Emma; and no surprise that a ten-year-old girl who was away from her home with the school for the first time, isolated from her friendship group, felt lonely and was missing her parents. Yet I must confess that the full depth of this feeling is not fully expressed here, Emma (and perhaps we) were holding back a little from the full extent of the emotion.

5.2 More stable and diffuse or intense and unstable friendships

There were differences in the intensity of friendships between young people, with some friendships being intense and often more conflictual, and other relationships being more relaxed friendships by association than expressive of a deep emotional reciprocity, and by implication, less conflictual and intense. There was a gendered patterning to these friendships, with girls tending to have more intense emotional relationships with higher levels of instability and boys claiming to have more friendships of association, which were less likely to fragment and fall apart. In general, boys did not talk in as much depth about their feelings in relation to friendships:

Clarence: I don't really have a best friend, I don't like, I don't really like saying oh you're a good, I don't really like saying oh you're a good friend but I like him more or something.
Int: Oh OK, but you're quite easy going all of you aren't you?
Adi: Yeah.
Clarence: Girls always have like friends' crises, don't they? Girls always have like friends' crises because they're all like oh she won't play with me.
Int: But you guys don't have friend crises, no?
Bevis: No.
Int: No OK.
Bevis: If Clarence doesn't play with me, I wouldn't care, I just play with Liam or Adi. I don't care if like Clarence wasn't here, because, well I would care . . .
(*All laugh and talk*)
Bevis: Well I would care, no sorry, sorry, sorry, I would care but . . .
Clarence: Yeah, but I have like loads of other friends, like this whole class is my friend.[12]

It is intriguing that even though these young people are claiming that they are not overly emotionally invested in their friendships, every child in this focus group took similar photos of the same group of young people – themselves. It is possible that they are performing a macho identity rather than expressing their genuine feelings towards their friendship, which was perhaps more important to them than they wanted to confess to each other.

Some boys were more open about discussing close and best friends: for instance, Conrad[13] mentions: "I'm friends with someone, just not, just, I've known him all my life – just one thing, he has ADHD and he's rather naughty", and later, Conrad identifies this friend as "*my best mate*". Similarly, in an interview which started off as an individual interview with Graham, but then became a paired interview when he requested his friend to join us, it was clear that Graham and Jason were very close friends,

> but he plays with mine and he always takes it off me. It's like, Jason comes to my house sometimes, and like all my stuff is like, his stuff, because we always share our stuff. And he says things like 'can I share that with you and say that it's mine as well?' and some of his stuff is mine as well. Like some of the stuff like my wrester, he wants it for his birthday, and so he likes to play with that. So it's like, I said "every time you come to my house, you can pretend it's yours and you can play with it the whole time." He can play with what he wants, and I've got loads of records.[14]

Some girls also discussed having different groups of friends to suit different purposes, with some close friends that provided emotional support and other friendships which were more of association or interest. Emilia outlines the distinction between different types of friendships, although other girls also emphasised that these friendships merged and the relationships shifted over time:

> I have two different groups, and Mel's in the one where I go and talk to them about like different things to this group, and this is like mostly the people that are in the group I talk to about like comic books and stuff! . . . Well I don't know, we kind of, we went to watch like The Avengers, like the Marvel movie and we've kind of gotten into it a bit more . . .[15]
>
> . . . so you kind of share everything, it's nice to have friends that you're not as close to because you don't feel like you have to tell them everything, and you don't feel emotionally dragged down! (*All laugh*).
>
> (Sabelle[16])

5.3 Dynamism and provisionality of friendships

The research as a whole was a short transect in time and space; it involved repeated meetings with the young people and different ways of approaching their worlds but was time-limited. On occasion we caught the dynamism of friendships as they were forged or as they broke apart. For instance, in Holt et al. (2017) we observed the emerging friendship of Beatrice and Shelly, two white British girls, in the coastal special school, despite their many differences.

Similarly, in the interview between Laura and Emma, we saw the emergence of their friendship, although also evident are the forces pulling the girls apart: Jennifer's research diaries give insight into the collapse of the friendship, in these various extracts of the rural primary school research diary:

> Laura and Emma had a falling out and wanted to talk to me about it but I suspect that Emma wanted to talk and Laura didn't really. Emma was

> having issues with the fact that it was never just the two of them and she always had to share Laura – I really had no answers about that, they should talk about it and see what compromises they could come to – I went back to look after the others as the noise levels had increased (drawn upwards arrow) and I felt I couldn't just leave them to it. E and L sat down and then I went back out. One of the cooks had gone across to see them because it looked like Emma was crying and upset. She bent down and chatted to them and came back out with a Twix for them. I went back to check they were OK and they said they weren't really and when I asked if there was anyone else to chat to, Emma had tears absolutely rolling down her cheeks. There was no time for "closure" before the end of break and every time I went to chat to them someone else came and interrupted [later] . . . Laura comes across to me and says about what Emma has been telling me, and it's not all true and actually she's been quite mean. I reassure her and she says it's good to play with other people and Laura said it's all Sally's fault, and that's not fair
>
> *[Later the same day, playtime].* So I went and spoke to the teacher about Emma and Laura and she rolled her eyes when I mentioned it, saying that Emma is a prime drama queen and that they had given them some time to sort it out. She said "sometimes it's better just to let them get on with it and they'll sort it out" . . .
>
> *[The next day, playtime]* Laura and Emma are separate – they've not been together all day. Both have found other children to hang round with. I asked Emma how she was earlier on and she said she was wearing the colours of South Africa, not referring to what went on yesterday.

The paired and small group interviews were semi-ethnographic, allowing the friendships to play out in the context of the interview discussion, and Jennifer wondered whether the research had perhaps precipitated the falling-out. I suggest that Jennifer is over-reflective here, and actually the friendship was precarious – something that was evident in the aforementioned interview. Nonetheless, this potential is an ethical question, and the conversations we had are presented here openly for critical reflection by the reader. It is intriguing that the teacher typecasts Emma as a 'prime drama queen', and I feel that at times as a researcher I (and we) have pulled back from engaging with the visceral emotions of these fractured friendships. How often do we, as adults, belittle young people's social experiences because it is too troubling and difficult to deal with the reality of the feelings of grief and loss that young people experience; albeit these feelings are often transitory as friendship groups are broken, reforged and reconfigured. On the other hand, adults' well-intentioned interventions can also be damaging (see Chapter 8), and often young people can 'sort it out themselves'.

Mahal and Jasmin from the coastal primary school were similarly close friends, just like sisters, as they stated earlier. Over the course of the research, we saw their friendship as a close and emotionally nurturing relationship, and then witnessed it fracturing in its exclusivity and closeness. Jasmin's adoptive mum talks about the friendship:

> So, she [Jasmin] tends to want to have one soul friend, and she did for a while actually, sort of then she got friendly with Mahal, I don't know. . . . And they were very good friends for a while. . . . Yeah, they were really good friends and then, then what happened, I think Mahal sort of decided that she wanted to also be quite friendly with Rita, so she took on another friend and Jasmin doesn't get on sort of with Rita. . . . So, but she doesn't have, I'd say she doesn't have a particularly, one definite friend. But it's really interesting so that a new child sort of became Mahal's best friend has completely upset the apple cart for a lot of children, you know it can really kind of thrust, you know, because all of a sudden they all kind of have to kind of shift around to try and you know who their [friends are].
>
> (Jasmin's adoptive mum)

Friendships are always provisional and dynamic, even if they endure in their tension between the need for emotional recognition and differentiation. It may be that Emma and Laura, and Jasmin and Mahal made friends again after Jennifer had left the school. These accounts are snapshots in time, and we do not know what happened after we left.

5.4 Reflections

This chapter has examined young people's social relationships in their visceral and emotional intensity, prioritising the words of young people. Young people have spoken in evocative terms about the importance of friendships and the emotional and social capital and recognition that they provide. It is also evident that friendships are fraught with power and with tensions. They are dynamic and shifting, characterised by who are not friends along with who are friends, and require work and effort to sustain. In the next chapter, I reflect on what more young people's friendships do in terms of reproducing enduring axes of power relations, including being a mediator in young people's relationship to school. The ontological need for recognition, which most (young) people experience means that young people as *contextual bodies/subjectivities/agencies* are always, already intersubjective and interdependent. Friendships of young people, then, provide emotional and social 'capital' and also position them within formative and generative frameworks of power that underpin subjection.

Notes

1 Mahal, British Bangladeshi working-class girl with labels of Social Emotional and Mental Health Difficulties (SEMHD), or with socio-emotional differences, coastal primary school, year five.
2 Some characteristics of the schools are provided in Appendix 1.

3 Jasmin, British mixed heritage black Caribbean and white girl with labels of Social Emotional and Mental Health Difficulties (SEMHD), or with socio-emotional differences, from working-class backgrounds, coastal primary school, year five.
4 Paavai, British Sri Lankan girl, unknown class origin, selective urban girls' high school, year nine.
5 Year nine focus group, selective urban girls' high school, with Saabira, British Indian; Karolina, white British; Erin, white British; Cora, British Chinese; and Samia, British Indian, all middle or upper middle class.
6 Annette, white British working-class girl with no SEND label, Church Street primary school, year four, talks about Rosie, who is a white British working-class girl living in foster care with a progressive visual and hearing impairment, Sara and Maya, whose demographic details I do not have. Most children in Church Street were white British and working-class.
7 Lucy, white British working-class girl with mild learning differences but not a statement of SEN, Rose Hill, year five.
8 Lindsay, white British working-class girl with physical impairment who used a wheelchair, Rose Hill, year five.
9 Emma and Laura are middle-class white British girls of the rural primary school, who were achieving well academically and who had no labels of SEND, year five.
10 Mary, middle-class white British girl, no SEND label, rural primary school. All the girls discussed were middle-class white British girls without SEND, year five.
11 White British middle-class girls with no SEND labels, rural primary school, year five.
12 Coastal primary school focus group with boys: Clarence white British, no SEND label; Bevis mixed heritage Black African and white British and SEND label for specific learning difference; Liam white British SEND label for a specific learning difference; and Adi British Indian and no SEND label, mostly from working-class or socially excluded backgrounds, coastal primary school, year five.
13 Conrad working-class white British boy without SEND label, coastal high school, year seven.
14 Graham, white British boy, from a poor/socially excluded background, with specific learning differences, Church Street, year five, with his friend Jason, white British working-class boy with no SEND label Church Street, year five.
15 Emilia, white British middle-class girl, selective girls' urban high school, year nine focus group.
16 Sabelle, British Turkish middle-class girl, selective girls' urban high school, year nine.

6 Young people as nodes of the intergenerational reproduction of enduring differences

(Re)producing subjectivities

In this chapter, I explore how friendships and social relationships are inherently powerful, with a generative and affirmative power of subjectification and recognition or more exclusionary forms of power, where other young people are isolated, excluded, marginalised or stigmatised. These, I argue, demonstrate socio-psychic geographies of 'othering' and of 'abjection' and 'distanciation'. These findings reflect upon the 'power' of young people's geographies to forge subjectivities that belong and do not belong to specific social groups or micro-geographies within the school space. The practices of 'including', 'excluding' and positioning are deliberate and reflect who young people identify and empathise more or less with; however, an affect, which is probably beyond conscious, is that these practices often follow lines of 'difference' and 'othering' around gender, socio-economic group and, especially, poverty, ability and bodily, emotional, mental and learning 'differences', and race/ethnicity. In addition, social and cultural capital intersect as those young people who have friends and good friends tend to have more positive views and engagement with school, although the broader socio-spatial contexts of schools also influence social relationships. In this way, young people, whilst being thinking, feeling, reflecting and agentic, are often 'nodes' in the reproduction of enduring differences. Of course, this is not all that they are.

6.1 Hierarchies of social groups and friendships

Young people's friendships and social groupings contained more or less subtle hierarchies between them, between the 'popular' and less popular young people. These hierarchies often reproduced broader axes of power around gender, class, race/ethnicity and dis/ability. The young people recognised hierarchies between different social groups, alongside drawing distinctions in their own friendship groups between 'close' and less close friendships. Friendships that extended beyond the school space were important signifiers and enablers of forging and cementing closer friendships.

DOI: 10.4324/9781003028161-6

6.1.1 Social groups and hierarchies: active (masculinity) – reinforcing gender and ableist norms

Eden[1] articulated that there is a distinct hierarchy between boys, and she stated:

> I always think there's a higher-class and lower-class; sort of all the boys that play football are in the higher class and the lower-class people play guns and that stuff.

Eden's analysis concurred with our own, and across the settings many boys played football or rugby (in Church Street rugby league[2]) and those who did not play these performative sports were relatively isolated or forged an alternative way of being in the playground, in mixed groups, either playing more exclusively with girls or exclusively with other boys, but at games other than sports. The importance of football is captured in Figure 6.1. For instance, Kasseem[3] emphasises:

> [I play with] Ahmed. Because Jack and Alistair are really good at playing football, and I don't like football, and I don't play it. And they play football, and Ahmed, he plays football, and sometimes he doesn't – usually he plays with me.

Figure 6.1 A football

Although Kasseem preferred not to play football, there was always the option of joining in a football game if there was no one else to play with. Nonetheless, the need to perform a strong and active physical masculine subjectivity, to be a boy in the highest echelons of children's social worlds, created barriers for boys whose mind-body-emotional differences made it difficult for them to perform this subjectivity. In addition, in schools with limited space to play, football or other ball games could physically dominate much of the space, meaning that those who did not want to or could not play football were marginalised both metaphorically and physically. This had gendered dimensions and intersected with normative ideas of masculine bodily competence. Whilst such patterning has been remarked upon previously, it was surprising to us that this continued in the new Millennium.

> As previously, the football match takes up the whole playground and those who aren't playing are relegated to the fringes of the playground – standing along the edges, sitting on the benches along the side, walking up and down the lines on the playground, or bunched in the corner of the playground. The football divides the playground on a somewhat gendered basis – no female pupils are playing (and as previously mentioned there aren't that many girls anyway), although a fair few of the boys are excluded from the football/choose not to play. The space is so limited that it would be hard to fit other activities in.
>
> (Research diary, coastal special school)

The differences we often (although not always) observed in boys' and girls' play, particularly in primary schools, are reflective of broader literature (Renold, 2005; Huuki and Renold, 2016; Pomerantz and Raby, 2020). Of course, any distinctions that we found in the friendships of boys and girls is not necessarily representative of a natural difference; any bodily distinction is always a process of iteration between 'nature' and socio-spatial contexts, given that young people (and indeed all people) are contextual bodies/subjectivities/agencies (see Chapter 3). It is also possible that our observations and analyses focused more upon the differences between the gendered/sexed subjectivities of the young people, as we have been socialised to do (Butler, 1997), rather than the continuity and similarity between them.

The gendered exclusivity of many social groups was part of a broader everyday politics of gender and sexuality which relied on a stereotypical gendered and heteronormative performance. These politics were troubled by young people who did not adhere to these dualistic norms. For instance, Violet[4] discussed the fact that they did not see themselves as female in a straightforward way, and although Violet experienced bullying for this, they had forged a 'gang' of friends around 'queer' gendered performances (see also Woolley, 2017).

Girls did also play football, rugby, bulldogs and other active games at times, and gendered identities were not fixedly masculine or feminine. Indeed, in Church Street and the special schools, mixed football and other active games were commonplace. For instance, in this excerpt from the rural special school research diary, a mixed group of boys and girls play football, and it is evident that the girls are more competent at playing than some of the boys:

> Alana[5] is now playing football with the boys – Andy[6] is not included but stands by the side waiting. Andy was included in a game involving carrying Bryn[7] around – well maybe not included but followed them around explaining what was going on to the dinner lady. The football game is ongoing – Andy is now on the field. When the ball is kicked towards him he cowers away from it. [Asha joins in] Asha[8] is obviously skilled and dribbles the ball, scores and celebrates vocally – hands in the air, screaming "yes". Two boys from other classes are in goal.
>
> (Research diary, rural special school, playtime, all students have moderate learning differences, and many have other labels)

Here, new lines are drawn, not around gender but around bodily performance. Indeed, the tendency to emphasise bodily performance was not exclusively masculine, with both boys and girls demonstrating high bodily competence, as Jennifer astutely notes:

> [It] is all about bodily competence – boys playing football or bulldogs and girls doing games which involve a rhyme and handstands, one of the girls says can I watch so I go across and watch. Very gendered games. Some girl doesn't want to do the handstands because she doesn't want to show her knickers but the other girls say it's OK, I don't mind, why do you? And the other girls promptly go and do some handstands, displaying their knickers. They decide it's OK for her to do the rhyme without doing the handstands. They are keen for me to watch.
>
> (Research diary, playtime in the rural primary school)

The importance of physical competence in young people's playground hierarchies meant that all young people could either be more included on the grounds of relative physical competence, whatever their gender, or relatively marginalised and isolated for their relative lack of physical competence. The marginalisation based on physical competence seemed to be most acute for young people with more hidden bodily differences, such as motor coordination differences, as discussed earlier. This tendency is evident in this game:

> I saw Noel who was with some other boys. They were all running and chasing each other, but Noel[9] was the slowest, and the others laughed at him a bit [on another occasion]. Then Joanna[10] and Rosie[11] started playing catch with Joanna's ball, and all of the other kids were watching them do that. Then Noel (who up until this point had been on his own, looking very lonely) asked Rosie if he could play. Joanna got really stroppy and said "It's my ball!" very negatively, and Noel looked really sad and sloped off.
>
> (Church Street research diary)

The young people often also reproduced normative, ableist expectations of bodily performance. These were often subtle and about motor control and performance, which isolated and marginalised young people who fell outside of these norms. In addition, young people with physical impairments who used wheelchairs or who were mobility impaired were never observed being included in these games. As suggested in the aforementioned quote, from Eden and by Kasseem, many young people who were marginalised within these performative cultures generated alternative social groupings around shared interests and a feeling of belonging.

6.1.2 Being funny, pretty, knowledgeable about music and popular

The young people often identified hierarchies of social groups within the school, with the 'popular' young people and the lower social orders (Morris-Roberts, 2004; Schäfer and Yarwood, 2008; Vanderbeck and Dunkley, 2004; Kustatscher, 2017; Thomas, 2009, 2011). Intriguingly, few of the young people identified themselves as being either in the most popular or the least popular groups. Eden, Mary and Jade[12] from the rural primary school talk evocatively about different social groups and their position on the social hierarchy:

> Like Evie's group – me and Emma used to call it the "Evie fan club" because like Lucy, Heather and Leanne they all followed her around, and Marge used to be part of that group but now she's like in with Ruth, Kiara and Heather. Erm and they, they, they're classed as the cool girls in our class and then I think the cool boys, I think, like boys who play football – I think that Adam and Jay because Adam and Jay are really funny.
>
> (Focus group talking about a group of middle-class white British children)

The most popular groups of girls were viewed as also being funny and fun, as Aashna[13] states:

> Yeah, like if you're, if, for example, like in a class there's always like the joker or the popular person, so if you're like for example the rock band person.

Although young people rarely emphasised that they were popular, Harriet[14] emphasised that she was a 'joker', and discussed her knowledge of contemporary music, which suggests that she is likely to have fallen into the popular category. In addition to their personality traits, the most popular girls were also often represented as 'pretty'. A senior teacher from the rural high school emphasised that Lily, a white British middle-class girl on the Autism Spectrum, had been voted prom queen:

> And the fact that one of our first ASD girls was voted prom queen last year . . . Yeah, and that wasn't a malicious, that wasn't a kind of sarcastic snide dig at Lily, that was genuinely recognition amongst kids who have had her in their classes for five years, and she wasn't necessarily the nicest of person, people to deal with at times, but recognition that hey that's brilliant you know.

The way in which the senior teacher goes on to discuss Lily speaks volumes about the way that adults in schools can reproduce negative and normative assumptions about young people with socio-emotional differences and/or who are on the Autism Spectrum. Lily had an entirely different take on the way she and the other girls were treated by boys in the school: "*But all the boys in school are kind of weird. . . . They always like call all the girls like babes and stuff, it's just weird, they're just weird people*". Also, both Lily and her mum emphasised the positive traits of her Autism Spectrum, such as a single-minded determination and competitiveness which had enabled her to perform highly in her academic studies whilst competing internationally in her sport.

6.1.3 Games and subtle power plays: performing ability, disability, race/ethnicity

Friendships are always imbued with power, yet these powerful social geographies were sometimes insidious, and subtle power plays centred around positionings within games or social groupings. Sometimes awareness of these powerful positionings emerged through dialogue with the young people. For instance, in our tracing of Emma and Laura's[15] friendship. Lindsay[16] and Lucy[17] spent much time together; however, they did not particularly identify each other as friends (see Section 5.1). Lucy drew upon Lindsay's physical dependence to enact power over her, whilst Lindsay emphasised her relative academic and social competence, which certainly within the context of Rose Hill, a relatively high-performing primary school with high norms of learning and behaviour, was more important than physical independence:

> Lucy and Lindsay asked me to sit with them. They were really pleased when I went in. We were all having a little chat. Lucy said to me "I have to stay here, because I have to look after Lindsay's care and needs", then she fussed over Lindsay, touching her like you would a kitten or something and then she put Lindsay's feet in her foot stands on her chair.
>
> (Research diary, Rose Hill, lunch time)

> Lindsay stayed in at play time [Lindsay had spent most winter playtimes inside, as did many of the other young people with physical impairments when it was cold, from a concern for their health; although this may have been a necessary precaution for their health, this did isolate this group of young people]. Lucy stayed in with Lindsay. I went and had a chat to them, again. They were looking at some old photos which were on display in the school. Lucy said to me "look, this is Lizzie." Lindsay asked Lucy "who is Lizzie." Lucy "a boy who used to go to this school." Lindsay "a boy! Lizzie's a boy." (And she had a good laugh at Lucy's expense).
>
> (Research diary, Rose Hill, playtime)

Another way that power played out amongst young people was the role they took in various games. For instance, some young people, particularly boys, who were less physically able, joined in football games, yet were on the periphery. In the rural special school Leon and Andy, discussed earlier, were simultaneously part of the football game and yet not fully included within the game:

> Leon[18] is on the edges of the football game and he goes to the male playground supervisor as soon as he comes out and gives him a half hug round his waist as the supervisor rubs him on the head and tousles his hair. They have a short conversation – Leon acts something out to him, then he wanders off and says "come on you Spurs" to some other children. Leon gets a ball and starts playing with it himself. He's now quite fluidly in and out of the game. Then he climbs up the netball hoop. He strides across the playground, chest puffed out – quite adult in his stance and manner. His trousers are a bit tight and his pocket is hanging out. He is holding on to the ball as the others stand round in conversation (maybe making up some rules) – Leon doesn't look hugely happy and there is some raised voice negotiation over the game and ball. Suddenly the ball is free and Leon is in goal at the other end of the playground – not yet having touched the ball. It hasn't come up his end at all. He looks very small and is standing still – just with his hands down his sides and rounded forward shoulders, then with one arm folded and the other holding his chin up. After a while he goes and gets the ball and does a goal kick intervening in the boys' discussion and static game. While it's cold some of the boys are in shorts and girls in summer dresses. The whistle goes and the children run across the playground to line up for lunch. Leon goes past and I again think he is very adult looking – maybe because of his slip-on shiny shoes which are quite different from the Clarks shoes most of the other boys are wearing (lace up, sensible). He feigns a kick as he goes past, as if in his mind he is still playing football, and despite the fact that he didn't really touch the football on the playground.
>
> (Research Diary, rural special school, lunchtime play)

In another example, which is also discussed in Holt and Bowlby (2019), some girls in the selective urban high school took on roles in games with racial undertones. In a game based on the Marvel films, such as *The Avengers* and *Thor*, a white middle-class girl, Emilia, played the biological son of Odin, Thor, who is a hereditary god, and Aashna, a middle-class first-generation Indian migrant, played the 'mixed race', part god, adopted son of Odin, Loki:

> It's like this comic book thing where it's like superheroes and the bad guy is Loki. . . . So like whenever we go and see these kinds of movies we always give a character to each of our friends. So I'm Loki, apparently, for no reason, just! And one of my friends [Emilia] is Thor, so we keep like teasing each other, being like oh I'm better than you . . . Loki's better than Thor.
>
> (Aashna, individual interview, urban selective high school)

On the surface, the girls were simply acting out the characters from a film that they had seen and enjoyed, playing with new possibilities and ways of being. It is perhaps interesting that Aashna emphasises that Loki is better than Thor. However, the casting of the characters seems problematic, with strong racial undertones. In the Marvel comics and films, Thor and Loki are reimagined from ancient Norse mythology, in which they are both gods (Carnes and Goren, 2022). Arnold (2011) points out that Thor in both the Marvel Comics and in Norse Mythology is unequivocally good and a god by birth – the biological son of Odin, who is the supreme god. By contrast, Loki is a changeling and shape-shifter – sometimes on the side of the gods and sometimes acting against them; Loki is not to be trusted. He is an adopted son of the god Odin, but he is from a race of giants, enemies of the gods, and is identified sometimes as a god and sometimes as a character akin to the Anti-Christ in Christian theology, precipitating the end of the world, Ragnarök. The plot lines of the Marvel comics draw upon what is known about the ancient Norse depictions of Loki, with Loki sometimes betraying Thor and the Avengers and sometimes supporting and fighting with them. Rodda (2022) identifies Loki as "mixed race", although the racial politics of Loki is not fully discussed in the book (Carnes and Goren, 2022), which focuses instead upon Loki's queer gender identity.

These differences were subtle, and they were perhaps operating beyond the direct consciousness of the girls, who were certainly not deliberately reinforcing racial and ethnic stereotypes. Perhaps I am overinterpreting innocent role-play between teenage girls who are experimenting with subjectivity. On the other hand, it is intriguing that the white girl plays the unproblematic, biological god, whereas the Indian girl plays the 'mixed race' and problematic god. As we questioned in Holt and Bowlby (2019), do these games give an insight into a "domain full of 'deeper' drives" (Philo and Parr, 2003: 285, cited in Davidson and Parr, 2014: 121), beyond, perhaps, the conscious, deliberate reflexivity of the girls?

In the final example I discuss an incident in which marginalisation was ostensibly (and problematically) homophobic and yet was a slur used to other and marginalise a girl who was both from a racial/ethnic minority and had learning differences:

> Lorna[19] was playing with Summer[20] and some other girls, including Aya. Summer and Lorna were holding hands, and then Aya[21] tried to hold Lorna's hand. Lorna said to Aya "Get off you lesbian, I'm not holding your hand. I'll hold Summer's, but not yours, because you are a lesbian."
>
> (Research diary, Rose Hill, playground at playtime)

Clearly, Lorna is being homophobic towards Aya; homophobia, and compulsory heterosexuality, was relatively commonplace in the school playgrounds, as has been discussed in other contexts (e.g., Bhana and Mayeza, 2016; Huuki and Renold, 2016), although some young people, such as Violet, actively contested this compulsory gendered normativity and heterosexuality. However, the intersections of subjectivity are complex, and Lorna is either deliberately or unconsciously drawing lines of difference around race/ethnicity and dis/ability in her actions towards Aya alongside sexuality; Lorna also othered and marginalised other girls from minority (or global majority) racial/ethnic backgrounds (see Section 6.2.1).

6.1.4 Extending beyond the school space: an important hierarchy of friendships

Distinctions were made between 'good' and 'best' and 'close' and less close friends. Friendships variously extended from school into home and other non-school spaces, or conversely from other non-school spaces, such as homes, leisure spaces or shared transport into the school. These were important signifiers of closer friendships and also important ways in which close friendships were forged. For younger children, having, and being invited to, birthday parties was an important signifier of being good friends. Most of the young people talked about parties they had been invited to and their own parties. In the following excerpt, which is typical, Sharon talks about parties she has attended and held:

Int: How exciting! Who else did you invite to your birthday party, anyway?
Sharon: Julie, Rosalind and Annette and Darcy.
Int: Have you ever been to anybody else's birthday party?
Sharon: Yeah. Rosalind, Julie's and Annette's. And I'm having a birthday party, I'm having a sleepover, when it's my birthday, because my mum says I'm too old to have a normal party.
Int: How many of you will you have for a sleep over?
Sharon: Just Julie, Rosalind, Annette and Witney, who is a girl on my street. She used to go to this school, but she didn't really like it. She didn't get on right well with all the people, so she left, and she went to XXX. She's my best friend. I'll have to sleep on the mattress, you know, like what you sunbathe on?[22]

A small number of young people in primary schools said they had not been invited to a party. Most of the young people who had not been invited to parties had SEND labels and often were also experiencing poverty and family challenges. Loretta explains that she is not invited to parties by her classmates:

Int: OK. Um, have you ever had a birthday party?
Loretta: (*shakes head.*)
Int: OK, have you ever been to anyone else's birthday party?
Loretta: You know James who's in the top Maths group. He's got three sisters called XXX and I've been to their party.
Int: How do you know them, do you know their family then?
Loretta: My mum's their mum's mate.
Int Have you ever been to anyone from your class's birthday party?
Loretta: No.[23]

In her interview, Loretta's classmate Sharon suggested that Loretta was sad, and claimed, "She's sad because she doesn't have anyone to play with". This contrasted with observations where Loretta was usually seen playing with other children, but perhaps does tie into the aforementioned discussion; Sharon goes on to claim that Loretta told her that she is sad because she has no one to play with, although Loretta did not mention this in her interview. Similarly, Nelson[24] seemed sad when he emphasised that he had never been to a friend's party:

Int: You can? Are you going to have one [a party]? Have you had one?
Nelson: Yes!
Int: Who did you invite?
Nelson: Liam, and Sebastian and . . .
Int: Who?
Nelson: Liam and Katie and Rebecca . . .
Int: Right, and have you ever been to anyone else's party from school.
Nelson: (*really sad*) No.
Int: No, never? *(In a sensitive and kind tone of voice, I think!)*
Nelson: No.
Int: Right, OK, I've got one last question for you Nelson.
Nelson: Um
Int: OK?
Nelson: Um
Int: Do you like ice-skating?
Nelson: Yeah!
Int: Have you been ice-skating?
Nelson: Yeah!

I am not quite sure why I asked Nelson about ice-skating, but I had run out of things to talk about, and I didn't want to leave him feeling negatively about his own subjectivity, and I had recently enjoyed ice-skating again for the first time in many years, so had been thinking about it. More critically, in the aforementioned excerpt, I think I pull back on engaging with the full extent of the emotion of the experiences Nelson is discussing, and I note how often in the interviews when the young people start talking about their feelings, we step back from this abyss. We are afraid of upsetting the children, but are we also afraid of how their emotions will make us feel?

I have reflected upon the ethics of asking young people these questions: could I have reified the importance of social relationships via my questioning? In subsequent research I have been careful to ask more open questions. I feel equivocal about this issue, and on balance I hope that any exclusions and marginalisations were already present and that I did not harm the young people through this questioning. I hope that any emotional harm was transient and certainly that this book and my wider research can promote positive change for young people. Nelson certainly seemed happy when he left, laughing at a jibe that he made about my age, which he had guessed at being much older than I was at the time, and at the experience I had shared of falling over. I do not think that his emotional pain was because I asked him about parties but rather because he had not been invited, still I could have been more careful in the questioning, and I endeavoured to be in subsequent research.

For older young people, going independently either into town or the city centre or friends' houses was important in forging deeper friendships. Most of the young people talked about leisure time spent independently with friends, when they would 'hang out' either at home or on street spaces and go to shopping centres. Out-of-school clubs tied to hobbies such as sport or drama were also important to the young people, and they were sites in which they could make friends who sometimes did not attend the same school. This could provide direct social capital, but also emotional capital, embodied as a sense of confidence in one's ability to make friends – particularly important for young people who were struggling with formal and/or informal elements of school (Holt et al., 2013).

For some girls attending the urban high school, a selective state girls' school, the pressures of the curriculum and high academic expectations, alongside some serious extra-curricular commitments (e.g., competitive sports), meant that time to meet up with friends outside school could be constrained. In this school, which some girls travelled over 40 miles to attend, the distances between girls' homes limited the amount of contact they could have outside school time. Despite these barriers the girls put much effort into scheduling sleepovers to spend time together outside school, as this group of girls emphasise:

Laura: And we have a lot of sleepovers.
All: Yeah.
Int: So you tend to see each other after school, like on evenings and weekends?
All: Yeah.
Int: How do you get, how do you travel to each other's house or place that you meet?
All: Bus.
Train.
Bus and train.
Car.
Int: Ah ah, your parents drive you there?
Patty: Well, it depends on how far away they are.
(All laugh).

(Focus group, selective urban girls' high school[25])

Those young people who were not permitted to go to sleepovers, to invite friends home or to do other informal out-of-school social activities could be relatively isolated from forging deeper friendships. Some young people with SEND labels were not permitted or not able to be independent due to either real accessibility issues or parents' concerns over their ability and competence to navigate public space independently. For instance, Holly[26] stated:

> What I want most is to go into town with my mates. My mum and dad won't let me. They don't think I can do it.

Transport was an important factor for young people being able to attend out-of-school leisure activities, both more formal organised clubs and more informal meetings between young people. In addition, transport was also an important space in which friendships could be made. For instance, the girls in the selective high school discussed friends they made on buses and trains. In addition, some young people in the special schools made friends with other young people in either adapted buses or taxis. However, transport was also a real constraint in enabling sociality outside of school. Some young people, with SEND labels, who travelled by taxi to school were able to be dropped off at a friend's house or picked up after a club. For other young people, changing their usual schedule of transport was seen to be absolutely impossible and there was no flexibility about arrangements; this made a significant difference to the young people's opportunities to forge friendships which extended beyond the school space. Parents' ability or willingness to pick young people up from activities, alongside their concerns about their child's ability to navigate social situations and spatial contexts independently, influenced whether young people could participate in leisure spaces.

Lucia[27] explained that she was not able to go to the local youth club because her mum was unable or unwilling to pick her up:

Int: That's your usual evening. Do you ever go to any clubs or anything? Like I don't know, do you go to youth club or anything like that in the evening?

Lucia: Youth club after school? . . . No . . . I go home every evening.

Int: Do you? Would you like to go to clubs or are you not bothered?

Lucia: I'm not very bothered really.

Int: You're not really bothered, OK. And what do you do at weekends?

Lucia: Anyway, my mum won't let me . . . . Go to after school clubs . . . No.

Int: Is it hard for her to pick you up or something maybe, if she's got to drive to get you; has she?

Lucia: Yeah.

Int: OK. But you don't know why she won't let you? OK. So if you really, really wanted to go do you think she'd let you or not? Like if it was really important to you.

Lucia: I really want to go but she still won't let me . . . I keep asking her and asking her and asking her every day.

From the Coastal Primary School, Jasmin's adoptive mum reported on how she believed that Mahal not being allowed to go to sleepovers and not inviting friends home meant that she was somewhat left out:

> I mean interestingly one of the difficulties I think with her friendship with Mahal was that it was very much kind of one sided in terms of, and I think that might be a cultural thing, that it was always us inviting sort of Mahal, and that often didn't kind of, she didn't get invites kind of back, and so it's quite hard, and also Mahal couldn't ever come and stay for sleepovers, so she kind of was left out when it came to parties and arranging sleepovers.
>
> (Jasmin's adoptive mum talking about Mahal and Jasmin, coastal primary school[28])

It is interesting to note that Jasmin's mum assumed that cultural differences prevented Mahal from holding or attending sleepovers, yet there may be more mundane reasons, as suggested by Eden in the last paragraph of this section.

Nevertheless, Aashna[29] and Paavai[30] from the selective girls' high school were clear that their parents would not allow them to sleep at other people's houses, with Aashna emphasising that this is "an Indian tradition"; rather than isolating the girls, this was negotiated by going along to the sleepover but going home at bedtime.

Other young people emphasised both the importance of sleepovers or going into town independently and yet that this was something that they did not do very often. For instance, Leroy[31] stated:

Leroy: I like, maybe like having a sleepover with all the boys and all, or having like, or like if, or having, going paintballing or . . .

Int: Yeah?

Leroy: Yeah.

Int: Are they things you get to do often, like . . . ? (interruption at door – chatting) Is it something you need to deal with or are you alright for ten minutes?

Leroy: No. not really . . . I'm going, I am booking, my mum's going to, when she's gone for her back pain, because she's slipped a disc.

Clearly, young people's broader social contexts, their family situations, how far they live from school, the transport they take to school and the level of flexibility in their transportation can influence their opportunities to develop and forge deeper friendships through socialising outside the school space. Sometimes these friendships can be constrained via mundane practices, as Eden[32] states:

> em, I'd invite more people round to my house, because we don't normally because my mum says not, em, not tidy enough.

Importantly, however, for some young people with labels of SEND, their socialities outside of school were severely curtailed by a sense of inflexibility in their transport, whereas others could be picked up at a later time from a club, school or even a friend's house. Surely this is a simple matter to change?

6.2 Mapping 'the same' and 'the other': gender, race/ethnicity and dis/ability

In the previous section, I have charted some of the subtle plays of power permeating young people's social relationships. Sometimes, however, the differentiations were stark indeed, with clear social geographies of inclusion and exclusion mapped upon playgrounds, and evident processes of identification and disidentification and even abjection wrought through socio-spatial practices and other performances, such as the language used (Valentine et al., 2009; Valentine and Sporton, 2009; see also Chapter 3). Often, these processes produced social geographies that reflected broader relations of intersecting axes of power, tied to gender, race/ethnicity, dis/ability and, rather than 'class' per se, the abjection of young people in poverty.

6.2.1 Exclusive friendships of reciprocity and recognition

Throughout the research, overall, young people tended to forge friendships with people with whom they identify and therefore social groups often had many characteristics in common. For instance, in many of the schools, gender-dominant groups were commonplace and young people of all ages in most contexts tended to form groups that comprised all or mostly boys or girls. This has already been evident in the aforementioned discussions. More subtly, the higher-status 'popular' groups of girls and boys tended to be gender-dominant or exclusive in most of the mainstream primary and high schools. There were exceptions. In Church Street primary school, most of the children played in mixed-gender groups, yet most young people discussed friends of the same gender. This patterning of the discussion of gender-exclusive 'friends' was pervasive, although there were mixed-gender friendships in all of the primary schools, with the exception of the rural primary school. In the rural and urban special schools mixed-gender groups were the norm, and they may have provided new 'lines of flight' to develop new ways of being in the world.

In the playground of Rose Hill primary school, I observed a gender and broadly ethnically exclusive (although there might have been ethnic differences of which I was unaware) group of girls, who mostly wore hijabs and whose families originated from the Middle East. Most of their parents were professionals or university researchers and academics who were working in the city, usually at the university, for a fixed period before returning to their home countries. This group of girls often came and talked to me in the playground, and the following quotes emphasise how one of the girls, Rhana,[33] reflected on her friendships, which were exclusively with other girls from the Middle East.

Int: OK, and who are your friends at school?

Rhana: Well there's um – Nadyia, and um, Aisha, and Farah, and Karyme and that's it really.

Int: And who is your best friend?

Rhana: I haven't really got a best friend. I had a best friend, but she left. She went back to her own country, um – Kuwait . . . She was in year 6, but when it was time to go to high school, her mum didn't want her to go, so she took her back to her own country.[later] That's one thing that I like about England is that I've got lots of friends, and – people from my country come here.

Int: So, are most of your friends from your country, or – not really?

Rhana: Well, they sort of come from my country, but not really Saudi Arabia, 'cause like Farah, she comes from Iraq, and um – Nadyia does, and maybe all of them, but I'm the only one who comes from Saudi Arabia. And there used to be one from Kuwait that was my best friend.

One day, I had been talking to Rhana and her group of friends, who often came up and talked to me in the playground, when I recalled this incident in the research diary:

> Another girl, Lorna[34] came up to me in the playground, just after Rhana and her friends had been talking to me and said ". . . watch out for those girls – leave them alone!" Before I could ask any more or challenge her, she had gone.
>
> (Research diary, Rose Hill, playtime)

This excerpt highlights that Rhana and her friends forged an exclusive group; however, it suggests that they may have been subject to racism and exclusionary practices by some of the other children. Note also that Lorna was involved in another incident outlined here which seemed to be stigmatising another girl around the lines of sexuality, but the other girl, Aya, was from an ethnic minority background. By contrast, however, Lorna was one of a group of girls frequently seen playing with Ali[35] (see Chapter 7).

Friendships were also forged around a shared experience of disability. In the special schools, groups were exclusively amongst young people with SEND labels, by default – there were no students without a SEND label in the schools. Of course, the experiences and labels of SEND were diverse amongst the young people, and the young people had various mind-body-emotional characteristics – a diversity which is hidden in the practice of segregating young people into special schools. Similarly, within mainstream schools with some kind of 'unit', young people often forged friendships within that unit that extended into other school spaces. In my research in a disinvested seaside town in the Southeast of England, I emphasise that the self-identification and mutual support of this group of young people provided a background to counteract experiences of bullying. In Seadale School, when Holly[36] shared her experiences of being bullied, Andrew[37] and Violet[38] offer their support:

Violet: That's right out of order [to bully people for being disabled].
Andrew: Yeah, that is out of order, yeah.
Violet: Which one. Is that Graham and Lucy . . . ?

I reflect that Violet might be suggesting that she will 'sort out' the perpetrators, as Violet had discussed earlier how her gang took violent action against homophobic bullying.

6.2.2 On not belonging, geographies of exclusion and isolation: being the other

Although most young people had friends, we observed many young people, especially those with labels of SEND and particularly those with socio-emotional differences and/or those on the Autism Spectrum, alone. Whilst this could signify that they are being excluded, and many times this seemed to be the case, when we talked to the young people, it is also important to note that it is not possible to read exclusion directly from young people being alone. Indeed, some young people preferred to be alone at times or much of the time, as discussed in the next section. Often, however, some young people were subject to exclusion, bullying and abjection, as discussed in the following – and indeed, all of the young people with SEND labels had experienced bullying at some point in their school career (in line with Chatzitheochari et al., 2015). Some young people experienced multiple and interconnected exclusions, and this is also discussed in the final sub-section.

6.2.2.1 Complex social geographies of being alone

In Church Street, I often observed Alfie[39] playing alone or sitting alone in the classroom, with little interaction with the other children. This is just one of many examples:

> I saw Alfie and for the whole of playtime he was standing on his own. He had no communication with anyone. He was standing near a girl, but they didn't have any communication.
>
> (Research diary, Church Street, playtime)

I did assume that the other children were isolating and ignoring Alfie; however, many of the children named Alfie as their friend. I never observed or had reported to me any negative comments or actions made towards Alfie – indeed, the children liked him and accepted his way of being in the world (see Chapters 1 and 7). This preference for one's own company, or certainly spending some time alone, was not unusual, especially amongst young people on the Autism Spectrum, as Alex states[40]:

> Well some teachers, like my maths teacher, she asked me do I prefer to be alone and stuff, and I said yeah, and I do, so that's why I'm like on my own in the row, so no one's sat next to me . . . I sort of just like my own company.

It is critical to try not to judge or normalise young people's social worlds; however, it is also important not to assume that all young people on the Autism Spectrum are the same or that all young people on the Autism Spectrum prefer to be alone; it is critical to engage and discuss with young people about their preferences.

6.2.2.2 Being isolated and excluded

In each school there were a small number of young people who were isolated or excluded for much or all of the time and were on the periphery or the outside of most of the social relationships. Many of these young people had labels of SEND. Often (although not always) these young people were also from poor backgrounds and rather than being broadly working-class were relatively poor within their contexts. This experience was shared by boys and girls. Nelson[41] was often alone, as in this research diary excerpt:

> Most of the children in the lunchroom are sitting with friends and chatting. Nelson was on his own and looked very unhappy. He didn't talk to anyone throughout the time he ate his lunch. Another boy from a different class was sitting in front of him, but they didn't talk to each other, or look at each other. They had no communication, either verbal or non-verbal.
>
> (Research diary, Rose Hill)

Lindsay[42] spent almost all winter playtimes inside. Although another child frequently stayed inside with her, Lindsay felt very unhappy and isolated at school. She claimed that she had no friends, and then, with prompting, one friend, and considered this to be since she did not like to go and play outside:

> I haven't got any, I haven't got any friends in the school . . . I play with Lucy, she's my friend, but she's my only friend . . . Um, I don't pay much attention to people who don't want to play with me. . . . I don't know. Maybe I just don't fit in. Lucy likes to stay in, the other kids like to go and play out.'
>
> (Lindsay, Rose Hill)

Although Lindsay and Lucy spent most playtimes together, their 'friendship' was certainly not unproblematic (see Section 6.1.3). In this situation, it is perhaps not surprising that Lindsay hesitated before mentioning Lucy as her friend. In the previous chapter, Lucy talks about her friends and does not mention Lindsay, and with prompting (because I had observed that they spent most playtimes together), Lindsay responded:

Int: Is Lindsay not your friend, Lindsay [surname]?
Lucy: Yeah, Lindsay. Do you want to know, who is disabled?

That Lucy had not included Lindsay, with whom she spent a great deal of time, in her list of friends because she had not thought to include disabled friends, demonstrates a stark level of othering by disability. Even though Lucy and Lindsay were observed frequently playing and spending their break times together, it seems that Lucy could not conceive of Lindsay as a friend because she was disabled.

There are many examples throughout the research of young people with SEND labels being relatively isolated, and it was a common experience. Lindsay did not report (and I did not observe) any overtly negative practices towards her; nonetheless, she felt that she did not belong. Lindsay's social experiences were evidently framed and constrained by the socio-spatial organisation of the school and the classrooms (see Chapter 8).

Sometimes, young people from Black and minority ethnic backgrounds were excluded and isolated; these patterns then reproduced broader socio-spatial patterns of inequality. For instance, Nicola was a white girl from a poor family who attended Church Street, from a Traveller background. She did not have any labels of SEND, although she did have some learning differences. She stated:

Nicola: Like, no my friends, I don't really have any friends.
Int: You don't have any friends?
Nicola: No . . . . Poor me. Yes, I've only got one friend in the whole entire universe.
Int: Who's that?
Nicola: Who's that? Not in the Universe but I've got one in this . . .
Int: Really good friend?
Nicola: Yeah.
Int: Who's that?
Nicola: I've got Ida.

Nicola's experiences of social isolation reflected the broader patterns of exclusion faced by the Traveller community, and Traveller children specifically (Hetherington et al., 2020). For many young people, relative isolation was also accompanied by more negative and active processes of stigmatisation and abjection, as discussed in the following.

In addition, Aadesh was often observed alone and was more isolated than the other young people in the ASD unit; he was a British Indian boy with complex labels of socio-emotional difference; he was often observed alone in break times and lunchtimes, as in this excerpt:

> Andy comes up and asks the adults present what they'll be doing in 20 years' time. I ask him what he'd like to be, and he says be an experienced police officer and have a mortgage. Aadesh[43] is sitting on his own. Leo and Aiden sit together. Andy sits down and starts eating.
>
> (Rural high school research diary, in the 'ASD' unit)

6.2.2.3 On being bullied, stigmatised and abjected: intersections of SEND and poverty

All of the young people with labels of SEND discussed their experiences of having been bullied; for the most part not in the school they currently attended. Most of the young people with labels/experiences of SEND had attended multiple schools, and experiencing bullying and being isolated in school by other young people, and even teachers, was a common reason that young people moved schools, often into segregated special schools. These pervasive experience of bullying supports previous research which has identified that young people with SEND labels experience higher levels of bullying than their counterparts without SEND labels/mind-body-emotional differences (Chatzitheochari et al., 2015).

More unusually, some young people emphasised that they were being bullied in their current setting. For instance, Aidan[44] emphasises:

> But other people, but it's quite good because some of the people I don't really like you know, like you could put it in the bully term, but . . . [they] are not really in my classes so that's good.

Holly emphasised that she was bullied because she is disabled. She also highlights that the school has been ineffective at sorting out the bullying:

Holly: Because this school's crap!
Int: Why is that?
Holly: Because, no because I get bullied and then no one sorts it out and then it ends up my mum having to come to school.
Int: Who do you get bullied by?
Holly: People in my class think that it's funny to take the mickey out of disabled people.[45]

Holly was one of the few young people who claimed that she was bullied directly because of her physical impairment(s); indeed, throughout the research, we saw or had reported little direct disablist name-calling applied to young people with bodily differences or evident impairments. We did, rarely, encounter disablist slurs; however, they were not applied to young people with bodily impairments.

By contrast, in a pervasively ableist environment, particularly in schools with high academic standards, negative terms were more frequently applied to young people with learning differences. As Jacob states here:

Jacob: They [the children on my table] say I'm stupid and stuff.
Int: They say that to you, do they? And have you told Mrs Wilson that?
Jacob: No.
Int: And what do you do when they say things like that to you?
Jacob: I ignore them, I try.[46]

Across the research, there are many examples such as this, including in the special schools; for instance, in the coastal special school during a lesson, the following record was kept in the research diary:

Adam[47]*:* Callum you're so smart, you're smarter than Sam.
Callum: Everyone is smarter than Sam. Even my bag is smarter than Sam.
Leon: And your bag can't talk.
Callum: That's the joke.

> Kyle tells Sam to shut up (he is talking to the LSA). Sam goes up to the board to do a question. Kyle shouts out 'dickhead'.
>
> Aron and Callum are chatting at the front. Adam and Ella are chatting. Sam isn't included in any conversations.

On another day:

> Aron and Sam had been messing around and the teacher had tried to get them to come and watch the demo – had said "we know who the doziest among us are". Callum had added "we know who the stupidest are" and pointed at Sam.
>
> (Coastal special school, in a Maths class)

Similarly, in contexts where high expectations of behaviour were pervasive, young people on the Autism Spectrum and/or those with labels of SEMHD were often negatively labelled and stigmatised by their peers, as exemplified in the following excerpt from the rural primary school:

> Ben engaged with banter about love, marriage and rings with one of the girls. "Hey you go and buy me a ring" [says the girl]. replies Ben: "I'll go and buy you a ring for a million pounds" . . . As I walk off towards the toilet I hear one of the girls saying to Ben: A million reason to dump you . . . 1) you're weird, 2) you're weird, 3) you're weird.[48]

Young people in schools with more pervasive poverty and with higher levels of learning challenges and socio-emotional difference, often tied to parental stress, mental ill-health and sometimes drug and alcohol abuse, were more accepting of a range of different behaviours, as indeed were the special schools. This reflected differing school ethos and cultures (see Chapter 8).

I find reading these accounts troubling and emotionally wrenching. The vast majority of the adults in schools cared about their young people and were committed to their education and well-being, and yet the very real suffering of these young people was pervasive and naturalised. I/we did not directly intervene on behalf of individual young people. Yet it was clear that isolation and stigmatisation were acceptable for these young people *because they were different*. Almost always the social issues these young people faced were accounted for by *their social skills* or other ways of being in the world and the issues were individualised within the mind-body-emotions of the young people, particularly those with SEND labels. This is a pervasive ableist attitude in schools and society which permits the enduring suffering of those with mind-body-emotional differences and naturalises this as expected and accepted. This is a pervasive social injustice in our society which needs tackling and addressing. In Chapter 9, I have some suggestions for positive action.

Some young people had multiple intersecting subjectivities that were subject to othering and exclusion. For instance, Andy in the rural special school was a white boy on the Autism Spectrum with a label of Moderate Learning Differences. He is adopted, and lives in a poor, lone-parent family; his adoptive mum is a widow, following the death of her husband. She doesn't work outside of the home and relies on benefits and was suffering with mental ill-health. Andy was obese and had a special diet and exercise programme in place which, whilst he continued to be obese, had led to significant weightloss. Andy claimed: "I hate school", because of the hard work he had to do, rather than citing issues with friends. When asked who his best friends were, he replied: "Well my, I, well it's hard to say because I don't like the children in my class". He was often isolated at school and was, as shown earlier in the discussion, often observed alone or playing on the periphery. Andy did not adhere to the relatively expansive norms of behaviour that pervaded his special school, where practices that might have been seen as out of place in other schools were accepted and commonplace. Other young people also often said that they did not like him, for instance, a girl in Andy's class, Alana (white British middle-class girl), stated: "Andy's not nice with me . . . He swears a lot . . . Yeah at me".

Similarly, Holly, introduced earlier, was a white British girl who uses a wheelchair and has visual impairments from the high school in a deprived seaside town, whose parents were both out of formal employment. She experienced intersecting exclusions and stigmatisations. Alongside facing disablist bullying by her peers, she states that she is excluded for not having the right kind of possessions: "They (young people in my class) say stuff like, 'Oh you get your bag from Oxfam' and I can't remember what else they say, but they say shit stuff anyway".

Many of the young people from relatively poor backgrounds (i.e., the poorest young people in the school – for instance, those who were extremely poor in Church Street and relatively poor in more middle-class contexts) were abjected. The research diaries and interviews are replete with examples, such as this comment from Loretta, a white British working-class girl on the Autism Spectrum, Church Street, about Noel, a white British boy from a poor family with many challenges and social services involvement, who often came to school dirty, and with non-specific learning differences and some motor coordination differences, at Church Street:

> Noel hasn't got any friends, because they all pick on him. They say he's got lurgies, but he hasn't.

Interestingly, Loretta and Noel often sat together in class and shared the same Learning Support Assistant (LSA). Perhaps Loretta was deflecting her own sense of isolation, which I commented upon earlier, by emphasising Noel's lack of friendships.

It is perhaps unsurprising that young people from poor and socially excluded backgrounds were abjected, particularly if they also had other intersecting axes of power that were marginalised in broader society, such as disability and particularly learning disability. Poverty and disability are widely negatively reproduced in an intensely neoliberal broader society as an individual failing (Tyler, 2009, 2013, 2020), and these young people's experiences and accounts both reflected and reproduced these broader, pervasive representations and performances. These tales of young people being excluded, isolated and abjected are stark and deeply upsetting and identify a pattern which educators should take seriously and address rather than overlook and naturalise. In the next chapter I highlight how some of these young people fought back, either verbally or tactically, and forged groups of mutual support and belonging to contest disablist or poor-shaming practices.

6.3 Friendships and interrelated impacts on formal education – intersecting social and cultural capital: friendships reproducing educational inequalities

The preceding sections of this chapter have emphasised that young people's social relationships are powerful; they reproduce enduring embodied inequalities and disadvantages through everyday practices of identification, disidentification, inclusion/exclusion, disavowal and abjection. Very often, as seen in this chapter, these lines are drawn around enduring and intersecting axes of difference that fracture society and reproduce inequalities and hierarchies; around gender, class, race/ethnicity and disability, and these axes of difference intersect. Importantly these friendships and young people's social capital connect with cultural capital. Friendships and social experiences of school are crucial to most young people's engagement with school:

> I went round last night and I woke up this morning, and I was thinking, I want to go to school today, because I want to see Jason.[49]

In addition, most young people talked about and took photos of friends in school, more than of anything else, although there were differences, with some young people taking photos of places, such as fields, pets and objects – particularly computers and gaming consoles. Figures 6.2 to 6.4 demonstrate some of the range of things

Figure 6.2 Friends

Figure 6.3 My pet

young people photographed, the photos are taken from themes in the study, by Iola Smith, for enhanced quality.

It was notable that young people who are marginalised and excluded in social relationships have a more negative engagement with and view of formal education, as exemplified by the quotes from Holly and Andy. Similarly, young people who experience difficulties in the formal aspects of school are more likely to also have negative social experiences, and many young people discussed having negative experiences at the hands of teachers, usually not in the school they were currently attending. For instance, Andy's adoptive mum (white boy on the Autism Spectrum with a label of Moderate Learning Differences) emphasised that he currently had a problem with his teacher. She had complained to the school, but no action was taken and she accepted this as one of the things that happens in life. Along with negative social experience with peers, bullying by teachers was another key reason for multiple school moves by young people with

Figure 6.4 My trampoline

SEND labels, and many young people mentioned having negative experiences with teachers in other schools. In what is by no means an isolated example, Alana (white British, middle-class girl with moderate learning differences, rural special school) stated:

Alana: I don't mind because I used to go in school, a local school. . . . But I was so, I didn't like it. . . . Yeah, because one of the teachers was mean to me.
Int: They were mean, were they? The pupils or the teachers?
Alana: No, one of the teachers.
Int: One of the teachers was mean? Oh that's horrible.
Alana: I hated her.
Int: You hated her?
Alana: Yeah.

Throughout our observations, we also sometimes observed negative practices from teachers towards young people; collating practices of inclusion and exclusion is the topic of another book, although some thought is given to this concern in Chapter 8. As an exemplar, however, the following excerpt shows how Aadesh (British Indian boy, unknown class background, with labels for Attention Deficit and Hyperactivity 'Disorder' and on the Autism Spectrum) was pervasively left out of young people's social relationships and was also 'othered' and marginalised by adults, even within the relatively accepting space of the ASD unit:

> They run a student of the week and have a box where people can put nominations in. There were a lot of blank pieces of paper in there and this really irritated the adults – they called it silliness and that it was annoying and unnecessary. They asked Aadesh if he had done it and he said he had. They went on and on and there was only one genuine nomination and about 20 pieces of blank paper. Aadesh was told he was silly and he was ruining it for everyone else.
>
> I noted Aadesh sitting next to me and being spotted fidgeting by the LSA. She made him go and stand beside her. He kept on fidgeting[50] so he was told to put his hands on his head. She kept on monitoring him and moving him back to the spot she had stood him on and putting his hands back on his head.
>
> (Research diary, rural high school 'ASD' Unit)

We had also seen Aadesh removed from a classroom for behaving in ways which were similar to the other young people, and on another occasion, he goes to the ASD unit after being removed from a classroom, for, as he states, arguing with other young people.

The ways in which formal and informal aspects of the school, the formal aspects of curricula and the informal sociality of young people, intersected and reproduced similar lines of difference really mattered to young people as beings and becomings. Young people who had difficult social relationships had more negative views about school. This could have an impact on their educational engagement, with a critical impact upon the cultural capital that they attain through both formal qualifications and informal ways of being in the world or an educated habitus; therefore, young people's social relationships are a far more important factor in explaining pervasive educational inequalities than is usually considered. Social and cultural capital intersect. These are overlapping and mutually reinforcing. Alongside all of the aspects in which young people can be variously marginalised and excluded from formal aspects of school, given pervasive ableism and normative curricular which are devised from bourgeois and ethnocentric perspectives, some young people are often further, or perhaps most especially, disengaged from formal education due to their problematic social relationships. This (absence of) social capital is far more influential, pivotal and central to young people's experiences of school than is often taken into consideration.

6.4 Reflections on the chapter

In this chapter, I have explored how, through their own powerful social practices and agencies, young people often act as nodes of the intergenerational reproduction of enduring differences: (re)producing subjectivities. Through their own everyday socialities and friendships – who they decided to play with, who they decided to be best friends with, who they invited to parties or to sleepover, and the nature of their games, who the popular young people are and who occupy more marginal or liminal positions – young people often reproduce broader axes of power. They are reproducing hierarchies which have little meaning beyond school spaces, such as around being funny or a knowledge about contemporary music, although these can be viewed as a kind of embodied cultural capital which might well have broader resonance. Often, subtly or in more obvious ways, young people often (re)produce enduring axes of power tied to dis/ability, poverty, gender, sexuality and race/ethnicity. In most cases the young people were not being overtly disablist, sexist, homophobic or racist, yet the effects of their social practices reproduced axes of difference on these grounds, usually in habitual and beyond-conscious ways.

It is critical to remember that these young people were lively, embodied, powerful, bodies/subjectivities/agencies, who actively forged new ways of being; whilst performances often reproduced axes of power, they were always provisional and an improvisation. There were always new potential lines of flight, potential ways to be otherwise and to challenge enduring ways of being and sedimented patterns of inequality. In the next chapter, I reflect on connections in time and space, exploring how repeated performances in time and just being together repeatedly doing similar things day after day, similar things which always have the chance of surprise, create a shared history between young people, forging new ways of being. I consider how lines of flight that challenge enduring differences open up new potential future ways of being which might impact on young people's future possibilities.

Notes

1. White British girl, rural primary school, no SEND label, year five.
2. Rugby league is played more often in the north of England and is associated with the working-classes, whereas Rugby Union is often played in private schools, elite state schools and also Wales.
3. Kasseem, middle-class British boy of Jordanian and Palestinian heritage, Rose Hill, with no SEND, year five, brother of Ali, who had a degenerative impairment and used a wheelchair, year four.
4. Violet, white British working-class young person with cystic fibrosis, high school in deprived coastal town, year nine.
5. Alana, white British middle-class girl, with moderate learning differences, rural special school, year eight or nine.
6. Andy, white British boy from a poor adoptive family experiencing many difficulties on the Autism Spectrum and with moderate learning differences, rural special school, year five.
7. Bryn, white British boy with moderate learning differences. We do not know if he had any additional labels or his class background, rural special school, year eight or nine.
8. Asha, white British middle-class girl with moderate learning differences, rural special school, year eight or nine.
9. Noel, white British boy from a poor family with many challenges and social services involvement, who often came to school dirty, and with non-specific learning differences and some motor coordination differences, Church Street, year five.
10. Joanna, white British working-class girl with Down's syndrome, Church Street, year five.
11. Rosie, white British working-class girl living in foster care after being removed for abuse and neglect from her birth mother, with a progressive visual and hearing impairment, Church Street, year five.
12. All middle-class white British girls with no learning differences or SEND labels, rural primary school, year five.

13 Aashna, middle-class first-generation Indian migrant, urban selective girls' high school, year nine.
14 Harriet, Black British upper-middle-class girl, urban selective girls' high school, year nine.
15 Both middle-class white British girls with no learning differences, rural primary school, year five.
16 Lindsay, white British working-class girl with physical impairment who used a wheelchair, Rose Hill, year five.
17 Lucy, white working-class girl with mild learning differences but not a statement of SEN, Rose Hill, year five.
18 Leon, white British boy with a SEMHD label, moderate learning and medical differences, rural special school, year four.
19 Lorna, white British working-class girl with some mild learning differences but not a statement of SEND, Rose Hill, year four.
20 Summer, white British girl, unknown class background, no learning differences, Rose Hill, year four.
21 Aya, British Asian girl with learning differences, unknown class background and a statement of SEND, Rose Hill, year four.
22 Sharon working-class white British girl with achondroplasia, who self-identified as a dwarf, with no other SEND but a 'statement', Church Street, year four.
23 Loretta, white British working-class girl on the Autism Spectrum, Church Street, year five.
24 Nelson, white British working-class boy with learning and communication differences, Rose Hill, year five.
25 Sabelle (British Turkish); Kaeya, (British Indian); Tina (white British); Patty (white British); Anya (white British) all middle- or upper-middle-class girls, selective girls' urban high school, year nine.
26 Holly, white British girl, who uses a wheelchair and has visual impairments, whose parents were both not in paid employment, Seadale High School, year nine.
27 Lucia, white British girl, unknown class background, with moderate learning differences, rural special school, year eight or nine.
28 Mahal, British Bangladeshi girl, and Jasmin, British mixed heritage black Caribbean and white girl, both with labels of Social Emotional and Mental Health Difficulties (SEMHD), or with socio-emotional differences, from working-class backgrounds, coastal primary school, year five.
29 Aashna, middle-class girl, first-generation Indian migrant, selective girls' high school, year nine.
30 Paavai, British Sri Lankan girl. Unknown class origin, selective girls' high school, year nine.
31 Leroy, white British working-class boy with specific learning difficulties in the coastal high school, year eight.
32 Eden, middle-class, white British girl, without any labels of SEND, rural primary school, year five.
33 Rhana, middle-class girl from Saudi Arabia whose parents were working at the university and planned to return to Saudi Arabia, Rose Hill, year five.
34 Lorna, a white British girl, without SEND labels, although she was experiencing some difficulties with learning, Rose Hill, year four.
35 Ali, middle-class British boy of Jordanian and Palestinian heritage, who had a degenerative physical impairment and used a wheelchair, Rose Hill, year four.
36 Holly, white British girl with visual impairments, who used a wheelchair, whose parents were both out of formal employment, high school in a deprived coastal town, year nine.
37 Andrew, white middle-class British boy with specific learning differences, high school in deprived coastal town, year nine.
38 Violet, white British working-class young person with cystic fibrosis, high school in deprived coastal town, year nine.

39 Alfie, white British working-class boy with visual impairment on the Autism Spectrum. Church Street, year five.
40 Alex, white British middle-class boy on the Autism Spectrum, rural high school, year ten.
41 Nelson, white British working-class boy with learning and communication differences, Rose Hill, year five.
42 Lindsay, White British working-class girl with physical impairments, Rose Hill, year five.
43 Aadesh, British Indian boy, unknown class background, with labels for Attention Deficit and Hyperactivity 'Disorder' and as on the Autism Spectrum, rural high school, year eight.
44 Aidan, white British middle-class boy on the Autism Spectrum, rural high school, year ten.
45 Holly, white British girl with visual impairments, who used a wheelchair, whose parents were both out of formal employment, high school in a deprived coastal town, year nine.
46 Jacob, white British working-class boy with some mild learning differences whose family were experiencing difficulties, Rose Hill, year four.
47 Adam, white British working-class boy on the Austism Spectrum; Callum, white British working-class boy, from a poor background on the Austism Spectrum; Leon, white British working-class boy from a poor background unknown label; Sam, white British working-class boy unknown label; Kyle, white British working-class boy from a poor background, with specific learning differences and a label of Social Emotional and Mental Health 'Difficulties'; Ella, white British working-class girl on the Austism Spectrum and with an ADHD label. Coastal high school, year ten.
48 White British girls, likely to be middle-class given the demographic of the school, although we did not know the girls, and Ben, a white boy with a label of SEMHD from a poor family with social services involvement, rural primary school, year five.
49 Graham, white British boy, from a poor/socially excluded background, with specific learning differences, Church Street, year five.
50 On what I hope is the final edit of this book, I am struck by the disablist nature of this behaviour from the LSA, given fidgeting is a common trait of people with ADHD – this is like forcing a child who uses a wheelchair to try to stand up.

7 Immersive geographies and imminent transformation

Young people's powerful socialities – the power to challenge and change enduring inequalities

In this chapter, I examine the ways in which a shared history and trajectory of being together through time and space (immersive geographies) provide opportunities to forge identities/subjectivities in different, new and empowering ways. I develop the argument that these moments of performing subjectivities provide radical potentials to transform the ways these differences and embodied inequalities are enacted in society, and indeed, in future societies. In the chapter, I reflect upon factors which came together to forge these immersive geographies with their imminent potential for transformation. Arguably, schools are specific immersive spaces, and provide unique potentials for transformation given the particular dynamism of young contextual bodies/subjectivities/agencies and the repeated, regularised routines of the school space/time. I argue the coming together of young people and things to make inter-embodied connections, in a particular place, through regular repetition and the circularity of time, provides a potential to develop a shared history or habitus, forging new connections and providing new lines of flight and emancipatory ways of being that can challenge and transform enduring axes of power, albeit in these specific moments in space and time. Yet these new lines of flight have potentials beyond the specific space/times in which they take place. These new ways of relating can become part of the embodied subjectivities of young people, taken forward as habitus, or a way of relating and being in the world with their trajectories to the future. In addition, these new ways of being and relating, and the contexts of their emergence, could have resonance beyond the immediate space/time of the school, through countertopographies.

7.1 Moments of joy, moments of love, moments of empathy and moments of fun

The research diaries are replete with moments of joyous connection between young people, in which their emergent and dynamic contextual bodies/subjectivities/agencies exceed the bounds of any individual body to connect across porous (interconnected) material and psychic boundaries to exceed their individual subjectivity and to become more. These are ecstatic geographies which escape the realms of usual grounded possibilities. Katz (2004: 257), drawing upon Walter Benjamin, ascribes

DOI: 10.4324/9781003028161-7

to children's play a "mimetic faculty . . . to provoke an alternative, oppositional, and even revolutionary imagination that can see in the same, something different". She goes on to point out that play reminds us that "what is given is always made up and can thus be made different" (ibid.). Here I go further, to argue that children's play and young people's social relationships make the world different every day. These ways of making the world differently might be fleeting or more enduring, but they do have very real possibilities for disrupting enduring inequalities and prejudices. It is this very real set of possibilities that underpins imperatives to educate young people with a range of differences together. In the following extracts, there are some of the joyous moments we encountered during the research in which young people exceed the monotony of the everyday to find joy, laughter and friendship. Through their connections they exceed their own limitations (you need more than one to play chase). The connections are with other young people, sometimes with adults; things (beanbags, tunnels, photographs, chairs, wheelchairs, footballs, grass, flowers) were also sometimes agentic in the games:

> John has some beanbags and is laughing and his Learning Support Assistant is following them into the classroom. John looks really happy. They go back up into the hall but both Lloyd and Toby are now in the hall with me, playing with beanbags. John comes back with some more beanbags and Lee is chasing him now. They are running and his Learning Support Assistant is following.
>
> (Coastal primary school research diary, the children involved had some SEND labels and were white and British and working-class)

John had the highest level of support for SEND for SEMHD. Teachers claimed that he had many problems and difficulties at home and that he was angry and could be violent. When the research started, John was at risk of exclusion from the school and the headteacher was "wondering whether mainstream school was right for John" (research diary). However, John was spending some time in a special school and continuing to spend four days a week at the coastal primary school, where he was beginning to be more included. As seen in the aforementioned passage, the children transform the possibilities of the classroom and the hall space as the norms of appropriate behaviour are challenged and expanded.

7.2 Transforming disability into ability: new horizons of being

7.2.1 Fighting back: forging political sensibilities and solidarity in the face of exclusion and abjection

Some young people who were subject to pervasive abjection (see Section 5.5.3) sought out contexts and connections that provided them with positive relations of reciprocity and helped them forge new political subjectivities and ways of being that resisted the abjections to which they were subjected. For instance, Holly fought back against the negative labelling and bullying she experienced, and forged a collective and critical disabled group with Violet and others:

Holly: No, they just like to take the Mickey out of me because they know that I'll retaliate and I will retaliate.
Int: What do you do? What do you do if you retaliate?
Holly: I like use my mouth back, my only weapon I've got.
Int: And what kinds of things do you say?
Violet: I don't think you'd like to know.
Holly: No, don't think you would.

Although Andy was pervasively excluded and isolated in school, and had difficulties in his home life, he had some good friends at a youth club he attended that was organised by the charity Mencap, and has also recently started to play football with a club specifically for young people with learning differences, which he enjoyed and which might provide the skills and confidence to be more included in his school in the future, as a kind of embodied emotional capital, or habitus (see Holt et al., 2013).

7.2.2 New horizons of being

In the following excerpts, I suggest that the bodily differences of the young people are transformed to create new horizons and possibilities of being which are resonant of "enabling geographies" (Chouinard et al., 2016):

> Lindsay was in a wheelchair that needed wheeling, rather than her usual motorised wheelchair, and Lucy was pushing her. There was another girl with them, who I don't know. It looked like Lindsay was deciding where she wanted to go, because she kept pointing. They were zooming around, and Lindsay looked very happy. They all looked very happy.
>
> (Research diary, Rose Hill, playtime)

This can be contrasted with the accounts of Lindsay in the previous chapter, where she was often isolated and felt that she does not belong. In this excerpt, of course, the space of the playground itself made a material difference. The sun was shining, the weather was warming up, and Lindsay[1] was allowed into the playground, rather than having to stay inside. Furthermore, Lindsay seemed to be in control. Despite the fact that she was physically dependent upon Lucy,[2] who was, unusually, pushing her in a manual wheelchair, I argue that the extract shows mutual interdependence and empathy between the two girls; this moment in space and time provides an insight into a critical destabilisation of dependency (see also Morrison, 2022). The notion of independence is often applied to being able to physically do things for oneself, and yet critical disability scholars have emphasised instead the need for self-determination and freedom of decision (Watson et al., 2004; Lid, 2015; Chouinard et al., 2016). Here, it would seem that Lindsay and Lucy were critically interdependent, their porous bodies connected with the wheelchair, and they became an assemblage of girl-wheelchair-girl (air, ground, other children, sunshine) with new possibilities for connection and action. Lindsay helped Lucy to decide where to go and Lucy helped Lindsay get there. They were joyous in their whizzing around the playground.

Unlike Lindsay, who had expressed her feeling of alienation and isolation, Ali[3] was usually right in the centre of things. He had lots of friends and good friends, was rarely alone and expressed that he was happy and enjoyed school. In the following vignette, Ali is (not atypically) the centre of attention, and he is clearly the recipient of some new sunglasses. The sunglasses are perhaps a symbol of status and are certainly of interest, and they are a critical part of the scene:

> Ali was surrounded by children and a Learning Support Assistant. They were all talking to him, and he was the centre of attention. He looked very happy. He was wearing some new ski glasses which he showed me, and which seemed to be the subject of general admiration.
>
> (Research diary, Rose Hill, lunchtime)

Although Ali did sometimes have to stay inside, for most of the winter he was outside playing with the other children. He was always with other children, often with mixed groups of girls and boys, and frequently with Ben,[4] whom he identified as his best friend. The memory of watching the game in the extract here, makes me smile, and it is visceral to me. It struck me at the time as opening up new possibilities and ways of being, expressed in Figure 7.1, which visualises the child-wheelchair-child-child-child-child assemblage.

> All of the children were on the tarmac. Ali was wheeling around in circles with four girls holding onto the back of his wheelchair. Ali seemed to be doing the wheeling and the girls holding on. All of the children involved were laughing and obviously enjoying themselves.
>
> (Research diary, Rose Hill, lunchtime play)

Figure 7.1 The child-wheelchair-child-child-child-child assemblage

Source: Amelie Smith

Later, one of the girls involved outlined the meaning of the game:

Lorna[5]: [I like playing] Tig, and this horsie game that Ali made up – it's well funny. You have to hold onto the back of his wheelchair and you go giddy-up (*giggles*).
Int: So, is Ali the horse, and you the rider?
Lorna: No, we're pretending to be the horsies and Ali's a carriage *(laughs)*.

I argue that the child-wheelchair-child-child-child-child assemblage becomes a *Body without Organs* (Deleuze and Guattari, 1988), which forges a new way of being in the world, with emergent "lines of flight" (ibid: 109) that present new possibilities of being. Away from enduring hierarchical categorisations of disabled and non-disabled, new ways of being emerge. Possibilities are opened up beyond the dyad of ability/disability and beyond the constraining ableist or disablist framework, which pervades schools and societies, that assumes a particular type of body and mind. The assemblage of four children and a wheelchair (Figure 7.1) provides new possibilities of being and of intersubjective recognition. The wheelchair, so often a symbol of dependency and even of disability in a pervasively ableist society, is transformed into an interconnected machine of possibilities for creative play.

The ableism that pervades the lives of young people with learning and socio-emotional differences is often more difficult to trace or to challenge in a social and educational context where learning ability and particular forms of socio-emotional expression are naturalised. As emphasised in the previous chapter, young people with learning and socio-emotional differences and/or on the Autism Spectrum were frequently denigrated with disablist terms which went unchecked, whereas negative slurs around physical impairment were never observed as being attached to young people with bodily differences. Nonetheless, there were also times when young people with socio-emotional differences or learning differences were fully included and connected in relationships of recognition and empathy. In the following, Estelle was taking a pivotal role in swinging the big swing, which was a large log swing that could swing up to around ten children. We see the children all exceeding the usual constraints of movement with the swing as they flew through the air:

> Estelle was there with a group of children on a big swing and was really engaged – standing up swinging the swing for the others with real joy and abandon. It was the first time I had seen her really joining in.
>
> (Research diary, coastal primary school, at the park after school)[6]

In the following, John[7] was playing a game with a group of other children in which the train trays and the tummy trains allow them to exceed the limitations of their bodies, as they sped down the hill as a child-tummy-train-child-tummy-train-child-tummy-train

Figure 7.2 The sloping field (photograph Jennifer Lea)

assemblage. As the teacher emphasises, this exceeding was only possible because of the sloping field, shown in Figure 7.2.

> John is out doing playground golden time and we both look at him while we talk. They have the tummy trains out (like 4 wheels on the bottom of a tray, which all connect up in a "train"). Mr Paddington told me that these had been banned at lunchtime because they were too dangerous and I remark that they would take the legs out from under you, and the teacher says "it is great! It's the only situation in which the playground being sloped is an advantage!", John is playing with 3 or 4 other children on the tummy trains – they join them up and roll down the sloping playground and it looks great.
>
> (Research diary, lesson time 'golden time', coastal primary school)

From the start of the research period to this moment in time, and then throughout the rest of the research period, we witnessed a transformation in John's positioning by adults and other children. At the outset, John was socially isolated and at risk of exclusion from the mainstream school; in a conversation with a teacher, it emerges that

> [o]ne of the teaching support assistants is working one on one with a boy called John. The head teacher tells me that they are working with John in this one-on-one way with an aim to get him back into the classroom. He has behavioural 'problems' and [the head teacher] tells me that she is not sure that it is right for John being placed in mainstream school.
>
> (Research diary, conversations with adults, coastal primary school)

This moment of joy is facilitated by the fact that John and his friends had received 'golden time' as a reward for good behaviour, highlighting the importance of connections between formal aspects of the school and young people's social relationships.

In the final example, Rosie and Joanna were celebrated for their talents and abilities, regarded by their peers as entertaining performers, exceeding the normal limitations of everyday practice:

> Rosie had asked me to see her singing her song, so I went out at playtime to see it. Joanna and Rosie[8] were dancing together and singing, and I watched, then a lot of the children crowded around to watch, and were very positive saying things like, "oh Rosie, you're such a good singer!"
>
> (Research diary, Church Street)

7.3 Racial, ethnic and religious engagements

In schools with a diversity of racial/ethnic backgrounds, most friendship groups were of mixed ethnicity (with some exceptions of course, as demonstrated in the previous chapter). In the aforementioned example, Ali is from a Jordanian and Palestinian background and he is at the centre of a group of friends who were all white British. Ali's best friend was Ben, a white British boy who came from a poor family which had intervention and support from social services.

In the following, the girls from the selective urban high school reflected on a question posed by the researcher as to whether your friends need to be 'like you'. What the researcher meant had not been specified, but the girls took diversity to represent race/ethnicity (certainly not class diversity) and saw such diversity as a resource to generate new collective possibilities. Such inter-racial/ethnic groups are precisely the kind of groups based on empathy and recognition that underpin moves to educate 'different' young people together in schools and which also underpin Rev Dr Martin Luther King's dream. I argue that these friendships are powerful and do indeed have the potential to be transformative to society:

Erin: I don't think . . . I don't think it matters. . . . race matters that much, I think . . .
Sada: No, she's white, I'm Asian, she's black, Chinese! all laugh and talk.
Sally: No I don't think it matters as long as like you get on with them and you can have fun, I don't think it really matters like how they look and . . . or their cultures.[9]

Paavai talked about her previous experience of bullying and racism, which she contrasts to her current inclusive school and friendship group:

> [I]n [my primary school] . . . it didn't really work out and I was bullied and . . . basically, I don't exactly remember but I think it was about like the way, things I eat, ate and . . . My culture . . . Yeah, and then that's why

> I needed a big change, it was, I was really like a bit scared going from that school to this school because I was a bit scared about . . . how people were going to be, but everyone was just like me here and . . . Yeah and it was really easy to just make friends.

The girls viewed their ethnically mixed friendship groups as a resource, and they were clearly learning to be global and cosmopolitan citizens who were educated about a variety of religious and ethnic ways of being in the world, as exemplified by these quotes:

> It broadens your mind really. . . . I went to a private school where most people there were Christian, and I hadn't really had much experience with other religions and things like that. But when I came here, because there are so many different types of people, we found out about so many different hobbies people have. And it's really quite interesting.
>
> (Emilia)[10]

> Yeah, there's quite a lot of diversity, even in our friend group there is a lot of like different backgrounds and stuff like that . . . you feel that when you come to this school you can relate to people and like you have someone you can like, who is sort of the same as you, and people who are different, and you can really get along with those kind of people. I mean like if, because I'm Hindu and I believe in some things which other people may not believe in, but I still get along with other people.
>
> (Paavai)[11]

Paavai's quote is expressive of the ways in which encounters between young people can facilitate connections, which "blow apart strata, cut roots, and make new connections . . . rhizome-root assemblages, with variable coefficients of deterritorialization" (Deleuze and Guattari, 1988: 17, cited in Holt and Philo, 2023). Or put more simply, perhaps, these relationships crosscut ensuring power relations embedded into 'arborescent', or seemingly fixed categories, such as race/ethnicity, destabilising the categories of ethnicity by emphasising the connections that defy whether racial/ethnic differences matter at that moment, except to explore the magic and mystery of other people's religious beliefs through a dialogue of intrigue, wonder and empathy. These relationships were specific to this context and emerged not through a chance or fleeting encounter but through an enduring repetition of seeing the same people and doing very similar things day in and day out together. This provided opportunities to develop a shared history and connection. These deep empathetic friendships have possibilities beyond the immediate context. For instance, these girls were academically high achieving and likely to move on to important positions in key professions where they are likely to wield influence. As they mature and develop into women in society, the sedimented and embodied histories of these encounters are still part of their subjectivities and a critical backdrop to future encounters. The girls might move on from these friendships, but these lively, critical, caring, respectful, engaged, dynamic, intelligent, hard-working, funny, fun and playful girls will be present within the women they become, and this is powerful indeed.[12]

7.4 Shared histories and regularities: geographies of repetition

> Alfie sits still for about a minute before he is rocking again.
>
> Then he is sitting very close to another boy, who doesn't seem to mind.
>
> Mr Kane explains to the children what they have to do, and there is a lot of general chat about the work, to Mr Kane but also to each other as the children get their things ready. The children seem very excited, happy and interested.
>
> Meanwhile, Alfie is continuing to sit with his head on his hands, then he rests his hands on his chin, then he is clapping again, then he studies his fingers very thoroughly, and then he is counting again.
>
> The other children just don't react to all this at all.
>
> Alfie continues to do as above, as the other children continue getting ready.
>
> All the children laugh about something and after a pause Alfie laughs really loudly.
>
> Then the children move to the tables, and Mr Kane says they will go to the tables depending on how smartly they are sitting.
>
> Then Mr Kane dismisses the children, saying things like, if you're wearing purple, you can go. Alfie is in the second half of children to go and line up, with "people with names beginning with A".
>
> (Research diary, Church Street)

This extract appeared in the introductory chapter, and it is repeated here because, of the approximately 1,000 pages of research diaries written between us, this extract stands out as emphasising the powerful nature of repeated encounters to forge new and expansive norms and ways of being together, and shared histories and trajectories between young people. To me, at the start of my research journey, Alfie's behaviour and practices were of note within the context of a mainstream school (see Chapter 1); however, almost without fail the other children just accepted Alfie's behaviour as part of their everyday lives. They did not comment, nor was there a sense that they were deliberately trying to avoid comment. There was no awkwardness. Alfie was just Alfie; he had many likeable qualities: he was a good friend, he often preferred to be alone, but he could join in with other friends if he wanted. He rocked and clapped and sometimes shouted out or laughed at times when no one else was shouting or laughing. Sometimes he touched other children on the carpet, and mostly his peers let him touch their hair or their shoulders, but sometimes they asked him to stop or moved his hands away – no one was ever angry; they just calmly moved his hands. It seemed that they accepted that this was a form of touch that helped Alfie given that he had a visual impairment. In the research there are many examples where a shared history and trajectory, a daily repetition of similar activities, facilitated knowing young people with mind-body-emotional differences as people, and helped to forge connections through what is after all a continuum rather than a dualism of difference. The headteacher of Church Street had an interesting insight:

> Our kids adore Anthony, Neal and Joanna, because they know them as Anthony, Neal and Joanna. They've been Anthony, Neal and Joanna since reception. If you, um, live, play and work with a child with Down's syndrome, they are not a cretin [*sic*], an idiot [*sic*], or whatever . . . and hopefully they'll take that out into their lives. 'Oh, that's someone like Neal not something to be frightened of.
>
> (Headteacher, Church Street)

Around the time of the moment of joy discussed earlier, with the tummy trains, John's[13] positioning with other children seems to shift. In the earlier period of the research, John was almost always observed alone or with adults, notably his LSA. Then he is often observed with other children, notably Megan[14] and Lachlan.[15] As time progressed and through the repeated circularity of time, as the children encountered each other many times whilst engaging in similar activities (but wherein there is always provisionality and the potential to make something new and different, to become something new and different), new friendships emerged. Indeed new possibilities and ways of being emerged, with John and his friends exceeding the potentials of themselves as bodies, by connecting with other bodies and other things as they hurtle down the hill on the tummy trains. In the research diary, Jennifer reflected on how John had changed throughout the research period:

> To me, it seems that he has made "progress" since when I first saw him on the day I looked round the school and Mr Paddington said he was under threat of exclusion, and Mrs [the head teacher] said that she wasn't sure he should be in a mainstream school. John has a completely different facial expression – he is smiling and his face is relaxed and this is quite different from seeing his face tense and frowning with a furrowed brow which was what I saw previously.

As stated earlier, John was also spending one day a week in a local special school, and this might be a context for extra support and nurturing; it is interesting that what appears as an exclusion might, paradoxically, facilitate inclusion. Indeed, in an earlier paper we have argued that the most effective schools at supporting young people with SEND are often those that are porous and connected hubs where the expertise, services and therapies from 'special' education can permeate the mainstream school space and where young people can move between these contexts to facilitate gaining an appropriate curriculum and support (Holt et al., 2019a).

7.5 Reflections: skipping space and time – transformative possibilities

In this chapter, I have strategically sought out and presented the positive moments of encounter between young people and highlighted their transformative possibilities. There are, of course, many more in the materials, and the examples are selective, and sometimes strategically so, in order to focus on moments when new potential ways of being seem imminent. In these moments, the young people exceed the constraints of lines of demarcation around identity/subjectivity characteristics which are always powerful and hierarchical, such as abled/disabled, gender; race/ethnicity and class, and notably poverty and coming from a family with troubles or issues. On the one hand, these are just specific moments in space and time caught in our observations or conversations. On the other hand, these were often enabled by young people coming together in a specific space for an enduring and repeated amount of time. They got to know each other and built up a shared history or collective memory, a deeper understanding of each other from a range of perspectives, which facilitates a knowledge of connections and identifications, and facilitates empathy; even for people who are perceived as different, that difference has positive attributes and can denote exceptional talents, perhaps, and is not always less or 'other', just different. The circular repetition of time in a particular (though always dynamic and performed) space/place is not incidental to these lines of flight, these ways of exceeding, these new potentialities.

Space and time exist in a mutual reiteration – and time, like space, does not consist of discrete moments, of separate stills or photographs along a linear timescale which are left behind as each new moment is created. As outlined in Chapter 3, young people are *contextual bodies/subjectivities/agencies*. Consequently, the embodied subjectivities they become are at the same time material and corporeal, but even this materiality or 'nature' – blood and bones and genes and personalities – emerge in a dialectic with their socio-spatial context. Young people's subjectivities and their habitus or habitual ways of being in the world are a constituent iteration of their histories and their present context and their material predispositions. Of course, these young people grow up and grow older; they change as they encounter new people and new contexts; and yet, these moments and memories are still there as part of their embodied subjectivity, in the same way that my childhood memories inflect my present (see Chapter 1). We do not know how these moments of transformation generated new individual and collective, habitual and seemingly natural ways of being; however, these were not fleeting moments that are then left behind; they still exist in the embodied histories and emerging subjectivities of the young people. There is much debate about the transformative potential of encounters and their ability to be scaled up (see also Chapter 4). How much more transformative are repeated encounters, circularly going around and around again, with slight variations and with moments of sheer wonder and joy? Moments that happened, perhaps repeatedly, or perhaps once, and then are gone, yet are always present within the memories – embodied, habitual, forgotten or retrieved by the young people, and maybe also the adults around them. These moments are part of the adults that these young people became. This has the potential to influence their future encounters, exceeding the specific space/time. In addition, these new ways of being and relating, and the contexts of their emergence could have resonance beyond the immediate space/time of the school, through countertopographies. This book is a countertopographical project, exploring new and potential ways of being. In the next chapter, I reflect upon some of the contexts of these new ways of being, also considering how immersive geographies are porous and connected to specific socio-spatial contexts which constrain (although do not determine) and enable particular horizons and potentials of young people.

Notes

1 Lindsay, white British working-class girl with physical impairment who used a wheelchair, Rose Hill, year five.
2 Lucy, white British working-class girl with mild learning differences but not a statement of SEN, Rose Hill, year five.
3 Ali, middle-class boy of Jordanian and Palestinian heritage from a middle-class family with a degenerative physical impairment, Rose Hill, year four.
4 Ben, white British boy from a poor background, whose family had some involvement with social services and who had mild learning differences and physical impairments, but no statement of SEND, year five.
5 Lorna, white British working-class girl with some mild learning differences, Rose Hill, year four.
6 Estelle, a white British girl with a label of Specific Learning Differences from a working-class background, coastal primary school, year five.
7 John, a white British boy who has learning, social and emotional differences from a poor background, coastal primary school, year two.
8 Rosie and Joanna, white British working-class girls with a progressive visual and hearing impairment, and Down's Syndrome, respectively, Church Street, year five.
9 Erin, white British; Sada, British Indian, Sally white British, all middle-class girls, urban selective girls' high school, year nine.
10 Emilia, white British middle-class girl, focus group, selective urban high school, year nine.
11 Paavai, British Sri Lankan girl, unknown class origin, urban high school focus group, year nine.
12 Some of their representations of class difference were, however, more troubling, as discussed in Holt and Bowlby (2019).
13 John, white British boy who has learning, social and emotional differences from a poor background, coastal primary school, year two.
14 Megan, white British girl with complex learning and physical differences tied to brain damage at birth, poor background, coastal primary school, year two.
15 Lachlan, white British boy with labels of social, emotional and communication differences, from a poor family with 'difficulties' and a family history of learning differences, coastal primary school, year five.

8 Constraining and enabling young people's power

Reflections on the social-spatial contexts of schools

The preceding chapters have foregrounded young people's powers. In this chapter, I more fully trace the contexts of the emergence of these powers and the limits and constraints on young people's horizons. Although this is not reductive, in this chapter I want to explore some examples of how the young people in the schools are positioned in ways that constrain and enable their social relationships, and consider how they, and the schools in which they were being taught, were positioned in wider socio-economic contexts. The lively ethnographies of the previous chapters, the wonderful and enchanting and sometimes troubling and difficult, troubled and traumatised, personalities and subjectivities of the last few chapters are not equally positioned in relation to their access to capitals (economic, social and cultural) and potential future horizons – from the aspirations for their education and future lives to the actual economic and educational opportunities in their area, these can be vastly different even at very small scales.

Schools are, in line with how Philo and Parr conceptualise institutions more generally "precarious geographical accomplishments" (Philo and Parr, 2000: 517); specific moments in space and time in which people and things come together in unique ways which are at the same time porous, connected to and pervaded by broader socio-spatial processes. These broader socio-spatial impulses include resources, messages from the media, political contexts – what I have labelled the 'special' and 'general' education institutions; the inherent and intensifying ableism of the general education institution; and the resources, capitals, trajectories and histories of all of the people in the school spaces, which is intimately connected to the places in which the schools are located, and their positioning with broader space/time. Where schools are – the areas in which they are located, the way they are positioned within wider institutional frameworks and resources of education, social support and health, and who they recruit as teachers, teaching assistants and leaders – is critical. This intersects with the young people's own subjectivities, histories and trajectories. This does position some young people as "waste" and others as "precious commodities" (Katz, 2018), although these positionings are not entirely dualistic, and young people and others have the potential to contest, challenge and transform this positioning.

DOI: 10.4324/9781003028161-8

Nonetheless, some young people are more likely to be wasted. This includes the young person with a label of moderate learning difficulties or social, emotional and mental health difficulties, in an unpopular school with declining roles and stressed teachers with substantial sick leave periods, leading to multiple disruptions to their education. This young person will be disproportionately likely to be abused, be subject to child sexual exploitation, be excluded, become involved in the criminal justice system, be unemployed, suffer both physical and mental ill-health and to have a life expectancy which is 16 years lower than average life expectancy for women and 14 years lower for men (Heslop and Glover, 2015). In UK society as elsewhere, we have whole cohorts of young people about whom we naturalise and accept that they will have less fulfilling lives than their peers and that their contributions to society will be limited. This is disablist, and it is also about class and where you grow up and what school you attend. It is also about race and ethnicity and gender. It is intersectional, so poor boys of some Black and minority/global majority heritage, and with generic labels of SEND are the most wasted. In this book, there is not the scope to contextualise all of the young people's experiences and trace all of the ways in which their lively personalities are constrained and enabled by the accident of the geographical and social context of their upbringing and schooling. The schools across the studies discussed were selected to be diverse, and the broader contexts are outlined in Chapter 2 and in Appendix One. To trace some of these contexts, I have chosen to focus in depth on two mainstream primary schools from the same, large urban, LA in the North of England. I then go on to tease out some specific themes from across the research, focusing upon poverty, disability, labels of SEND and intersecting exclusions. In this chapter, I prioritise the voices of teachers, parents and young people in order to consider the interweaving of their experiences with the broader socio-spatial and institutional positioning of their schools and their families. I discuss some of the gritty reality of young people's lives, including poverty, social, economic and cultural exclusion at all levels, disability, disablism and abuse and neglect. This is not to be gratuitous but to highlight the very real challenges faced by some young people. First, I reflect on how the geographies in schools constrain and enable young people's sociality.

8.1 The geographies *in* schools – some reflections

The social and spatial context of the schools themselves was influential in facilitating or limiting young people's social relationships, and a key factor in how these social relationships forged social capital – the ways in which young people's friendships were positioned in relation to broader axes of power and resources. In this book, I have placed young people's own social relationships at the centre, and this is a deliberate conceptual and political act, given that their own agencies are so often overlooked. Nonetheless, the broader contexts of schools are critical. It is probably also the subject of another book, rather than this one, and deserves a focus of its own. It is helpful, perhaps, to reflect that all of the schools were ableist, teaching both consciously and unconsciously within norms and expectations of learning, behaviour and bodily development which, critically, dis-able those who fall outside of (and specifically below) these norms and expectations (Holt, 2004). However, the schools were variously inclusive and provided young people more or fewer opportunities to develop social and cultural capital and to connect with a diversity of other young people.

A key question in relation to the school-level education of young people with SEND is whether they have the best opportunities to thrive and to reach their potential in segregated special or mainstream schools. After all of the research I have conducted and read, and despite all of the people I have talked to, I cannot answer this question. This blunt dichotomy conceals the real question, which is: in which contexts do young people thrive and learn and have the greatest opportunities to forge meaningful friendships and achieve to their highest potential? This is about appropriate resources, support and knowledge, high expectations, the removal of barriers to socialising. The special school in the coastal area provided greater opportunities for its young people to thrive, achieve and socialise with others, both in that school and connected mainstream schools than was often the case in mainstream schools. In most of the mainstream schools, young people were in units or spent time sitting alone at the back of the class with a Learning Support Assistant or often in the corridor or other liminal spaces (see also Webster and Blatchford, 2015). By contrast, the special school in the rural area was isolated and inward facing, and the adults had low expectations of young people's futures and potentials, although the staff were kind and caring. The school was a container both for SEND and poverty.

Fully grappling with what is an inclusive school is certainly a question for another book; however, some thoughts on how the socio-spatial contexts of schools can enable and facilitate or constrain and limit young people's sociality are pertinent here. There are within the pages of this book many moments when young people's friendships are nurtured; for instance, the golden time tummy train play (see Section 6.2.2) and adults picked Ali up so that he could sit on the mat with the other young people. Without a doubt, Alfie's inclusion in his class and with his peers was supported by his teacher and the other adults and their calm and thoughtful approach to him. Teachers and other adults often dedicated time and energy to clubs where they shared their own passion with their students, forging spaces of association where young people could connect over a shared interest, providing a space of potential to challenge and overcome enduring differences and hierarchies.

It is evident from the pages of this book that young people's social relationships are foundational to their experience of school, in ways which are often overlooked or at best underplayed by adults. Young people need the time and space to have encounters where they can make connections, sometimes through 'difference' and sometimes perhaps around a coalescing identity characteristic – notwithstanding that all young people's subjectivities are intersecting. Some young people are more vulnerable to being labelled, excluded, marginalised, bullied and stigmatised in schools because of their intersecting mind-body-emotional difference and the labels that might be attached to them. All too often, adults within schools naturalise and assume that such young people will be bullied because they are different and link the bullying and the difference in a causal way. They then fail to take seriously and address young people's social isolation or exclusion, and their experience of othering. Even more frequent is a failure to realise that nurturing friendships and providing scope for young people, especially those who might be isolated or vulnerable to exclusion, to encounter others in a context of mutual respect and interest, tied to shared interests, with the potential to build connections based on identification and empathy, is essential.

Nonetheless, often adults within schools limited and constrained young people's sociality. For instance, young people were often excluded from the playground during playtime or break. Take for instance Lindsay,[1] whom we have met; she was often excluded from the playground. Generally, excluding young people from playgrounds (which are often constructed as a positive space for/by young people) is a spatialised disciplining strategy for those young people who have not conformed to expectations of behaviour. However, in Rose Hill, Lindsay and, to a lesser extent, Ali and other young people with bodily differences were frequently excluded from this space for health-related reasons. This constructed young disabled people as 'sick' (a component of the individual tragedy model of disability (Morris, 1991; see also Watson, 2012)) and restricted these young people's opportunities to build social relationships and construct informal cultures. In other school contexts, young people with socio-emotional differences and those with learning differences were often (in the case of the former) and sometimes (in the case of the latter) excluded from play or breaktime due to not conforming to behaviour expectations or not having competed set work.

We have already met Noel[2] from Church Street, who often had problematic relationships with his peers; his Learning Support Assistant (LSA) reflects on the fact that he is often kept in at playtimes:

> And some children, Noel, for example, is really slow at writing, desperate slow at writing, so he could do with it written out in paragraphs and just putting odd words in, or phrases, or something, but no, he has to write it out, and he don't get the work completed, so that it comes to the end of the lesson, and he stays in at dinner-time to complete the work.
>
> (Ms. Miller[3])

It is clear here that the lack of an inclusive or even differentiated curriculum is impacting on both Noel's experiences of learning and his social experiences. In many schools, young people were excluded from break or playtime on account of their behaviour, even when that behaviour was tied to a label or an experience of SEND and part of their mind-body-emotional way of being in the world. However, in the rural mainstream high school, young people with socio-emotional differences or who were on the Autism Spectrum were subject to a more positive restorative justice approach, which provides a positive exemplar of how challenging behaviour can be addressed in a more inclusive way (Lea et al., 2015).

Many friendship groups reflected the spatial organisation of the school, with many, particularly in primary school, reflecting the organisation of classes, and being in class together, was a key relationship of association that permitted encounters and the ability to make friends. Therefore, the actual material arrangement of classrooms and where young people sat within them in relation to their peers and adults, and which young people they sat with, had a significant influence on young people's ability to forge incidental social relationships in class that then extended into the playground. Despite this, many young people with labels of SEND were isolated from their peers and/or were given little choice over which peers to sit with. In most of the mainstream schools, the key resource for young people was a specific named adult, and this was tied to the funding regime. How these adults were deployed differed; although in most schools the adult was (almost) physically attached to a young person, the young person's learning was filtered through this adult and their social relationships often constrained (see Figure 8.1). In most schools it was possible to identify the young person with a

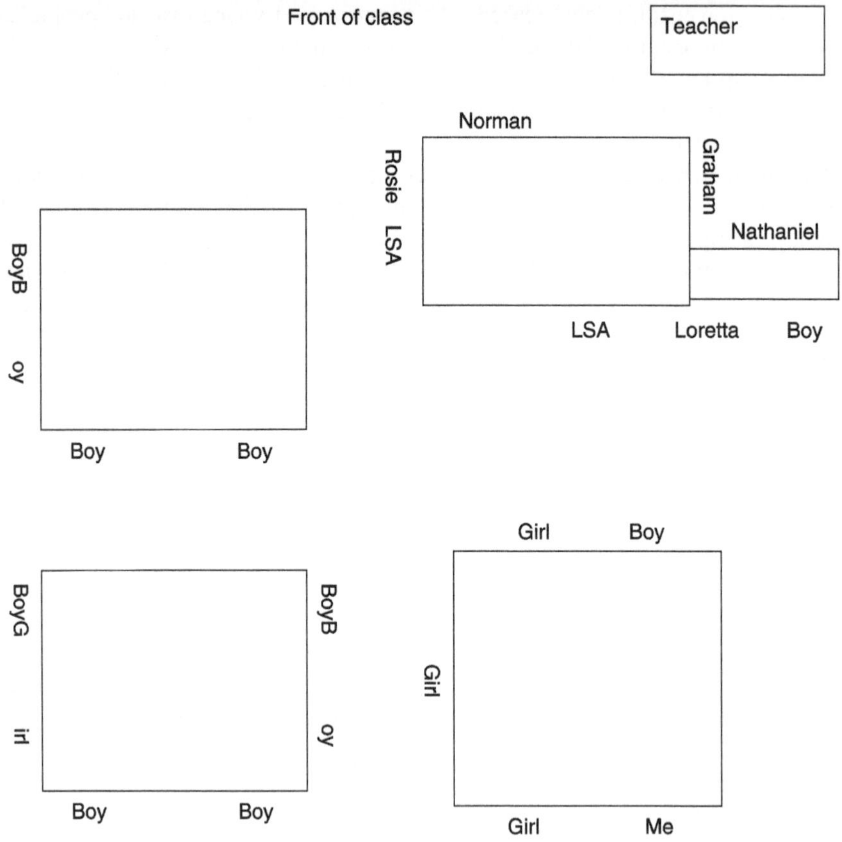

Figure 8.1 Plan of 'bottom set' numeracy, Church Street

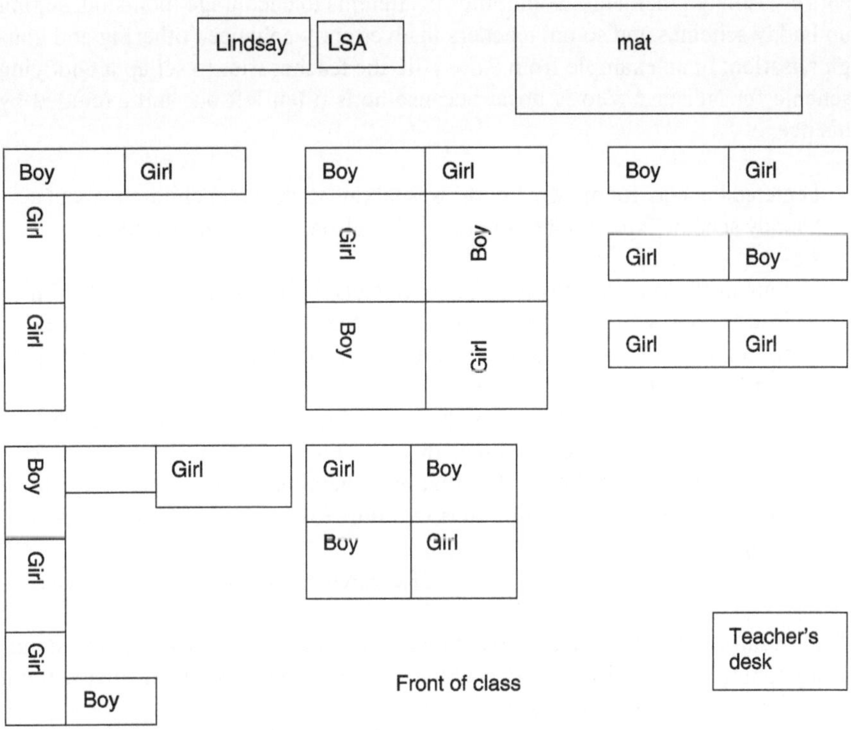

Figure 8.2 Seating plan 'top set' English Year Five

SEND label by the adult who sat next to them. For some young people with bodily differences, the classroom arrangement was based on expediency and this could lead to them being isolated or left out. This was the case for Lindsay, as can be seen in Figure 8.2 and the following quotes.

> Lindsay's LSA brings her in (after the other children) and helps her to get her things out. Lindsay sits in her wheelchair at the back of the class, away from the other children. . . . Lindsay does not interact with any other child, but there is some interaction between her and the LSA. . . . At the end of the lesson, Ms. Richards informs me that all of the other children seem to get on fine with Lindsay, but that sometimes they forget her when they hand out the folders.
>
> (Research diary: year five middle set Maths, Rose Hill)

Beyond these unintended consequences of expediency, some teachers actively excluded and stigmatised young people; some young people have already discussed their experiences of having been bullied by teachers, and these were not entirely isolated incidents. Sometimes in attempts to encourage inclusion, setting up buddy schemes and so on, teachers inadvertently enhanced othering and marginalisation: in an example from Rose Hill, the teacher tries to set up a buddying scheme for Nelson,[4] who is upset because he is often left out and excluded by his peers:

> I entered the classroom. Mr. Taylor[5] was talking to the other children, about a "buddy scheme" for children to play with Nelson. He said things like:
>
> > I am not asking you to only play with Nelson, just to try to include him in whatever you are doing. Because he feels left out, and then he tries to join in your games, and I know that is a pain for you. I know Nelson can be a pain sometimes, so in that case, you can tell the teacher. If he was being really silly, which I know he can be, just find a teacher or a dinner lady, and tell them that you are supposed to be Nelson's buddy for this lunchtime, but that he is being silly. You don't have to play with him if he is going to ruin things for you. I don't want to ruin your lunch times.
>
> (Research diary, Rose Hill, form time)

This example of a teacher directly intervening in children's friendships stands as a warning against directly intervening in young people's friendships without considerable reflection and care. No doubt Mr Taylor was trying to be kind to Nelson (who was sad, close to tears, and sitting in the corridor with his LSA whilst this conversation was being held). Nonetheless, it might be advisable to have taken a small number of children aside to ask them to buddy up with Nelson or, even better, to try to create opportunities for socialising around shared interests or something at which Nelson excels. Of course, Mr Taylor did care; he was (in my view) a caring and kind teacher who tried his best, who was working under constraints with a large class and high expectations of academic achievement. Any moment such as this needs to be positioned within wider constraints and contexts of schools, and I am certainly wary of 'blaming' teachers (except in cases of direct and evident bullying). Teaching is a critical, difficult and under-valued profession, and although I am imploring teachers to take more account of young people's socialities, particularly for those more at risk of being marginalised, isolated, excluded and stigmatised, I am aware of the pressures of the profession as starkly expressed in the recent case of Ruth Perry, who took her own life following a negative Ofsted inspection. This isolated example is demonstrative of a wider pernicious regime of under-funding, over-surveillance and expectation. Schools are, however, similarly connected by these institutional and governmental pressures and yet diversely positioned in relation both to these institutional processes and broader socio-spatial contexts, as discussed in the following.

8.2 Porous school geographies: schools' connections to the places in which they are located

The broader social and spatial contexts in which the schools were positioned, and which pervaded their porous boundaries as people, things, resources, ideas and so on, circulated between schools and spaces beyond the school at a variety of intersecting scales, were critical. These broader spatial contexts were important factors in both available resources and the capitals to which young people's social relationships gave them access. It is beyond the scope of this book to discuss in depth all of the contexts of all of the schools in the research; however, I discuss in depth two primary schools (Church Street and Rose Hill) and then go on to consider some generic points about context that apply to a broader range of schools. Both schools had a consciously reflected ethos of 'inclusion' for young people with a wide range of mind-body-emotional characteristics, and most teachers, other adults and the senior leadership teams emphasised the importance of promoting acceptance and inclusion of all young people with diverse characteristics. Neither school was fully inclusive of the entire range of bodily-mental-emotional diversity, and limits were set on the kinds of mind-body-emotional characteristics that were permitted within the school. In Church Street, limits were set around behaviour, and in Rose Hill, it was argued that they were unable to cater for complex medical conditions. In both schools, however, teachers held the view that inclusion of young disabled people was critical, both for their enhanced opportunities and education and to 'educate' young people as the future of society to be accepting of all kinds of mind-body-emotional states. As Mr Parker, the headteacher at Rose Hill, states:

> Um, really at the end of the day it's more reflective of society. You have a choice to make. You either have people who are regarded separate and put in special places, or you try to have an – inclusive society. So, therefore an inclusive society needs inclusive schools.
>
> (Mr Parker[6])

8.3 Church Street – a resource-limited school, with falling roles, in an almost exclusively, white, edge of city housing estate, with high levels of multiple deprivation

Church Street is a large primary school, with 385 students, in the inner suburbs of the city in the North of England, which is one of the most deprived wards in England as measured by Indices of Multiple Deprivation. It comprised 1940s and 1950s almost exclusively semi-detached, and 1950s terraced brick-built council housing, with gardens, which was almost exclusively, and relatively unusually in the UK, socially rented from the city council. Therefore, the housing was relatively affordable and of good quality.

The school had a high proportion of children with SEND, including children with various physical, learning and sensory disabilities, and one-third of children experienced some kind of SEND, although only 4.5% of young people had 'Statements'. This disparity between the number of students with SEND and those who receive the extra funding, resources and powers of a Statement compounded the school's financial situation and has been identified as an issue in other, similar schools (Ruth Lupton et al., 2010). Church Street had no extra resources to support children with SEND, except those provided by the children's Statements. A high proportion of students in Church Street came from socially excluded backgrounds. One-third of the pupils had free school meals, and many of the children came from families that are experiencing social and economic disadvantages. Levels of 'attainment' in Church Street were low. Church Street ranked close to bottom of schools in the LA Key Stage Two 'Standard Assessment Tests (SATs)'. The Special Educational Needs and Disabilities Co-ordinator (SENDCo) at Church Street claimed that:

> SATs and league tables are cruel – because they punish schools that have got a high level of SEN and they punish kids with SEN.
>
> (Ms. Gregson[7])

This LA had recently been deemed as failing by Ofsted and a then innovative corporate new LA had been introduced. At the time of the research there was a surplus of primary school places, and the school was subject to declining student numbers and financial pressure, and staff in Church Street felt that the school was at threat of closure.

A variety of explanations for the financial situation and the school's low academic performance were suggested, which included both external pressures and school-based factors. Perhaps unsurprisingly, school leaders focused on factors beyond the control of the leadership, whereas some other teachers (including one who had left the school and now worked in another case-study school, Rose Hill) also pointed to failures of leadership and mismanagement. Factors raised included the SEN process and a disparity between funding and Statement requirements (Ms. Gregson), the cost of vandalism, staff on long-term sick and supply teachers (Ms. Gregson, Ms. Mason, Ms. Massey[8]) and high wages commanded by senior staff (Ms. Trim[9]). Falling rolls were contended to compound existing financial difficulties. The school was considered to be a candidate for closure or merger, as it was unable to compete in the education market (Ms. Trim, Ms. Gregson[10]). A recent internet search shows that the school is still open. In the following, the headteacher and other teachers from the school give voice to the broader context of the school and the reasons for its interconnected challenges, focusing on its location within a deprived inner suburb.

8.3.1 Location in a deprived inner-suburb post–World War II social rented housing estate

> This area has got a really bad reputation. Well, it's South of the river isn't it, and you don't want to go there. There is a real divide in this city North and South of the

> River, but this suburb has got a really bad reputation. Some people who work here are ashamed of working here, but I'm not. I really love the kids and the parents. The things those kids have to contend with day after day and yet they still come to school, and work hard. They're incredible. The parents, most of them really struggle to do their best by their kids, although sometimes their idea of what's best for the kids isn't the same as ours. There are really strong kinship networks here, and most people here have been here for generations. People are proud to be from this suburb. When I first came here, I was really worried, but as I drove through, the thing I noticed was the gardens. People really look after their gardens. They are proud of where they live, and lots of them have lived in the same house for ages. I wouldn't hesitate to up sticks and move around here. Except for that I am only ten minutes away from [a beautiful natural landmark in the countryside to the affluent North of the city].
>
> (Ms. Massey[11])

Not only was Church Street primary school located in one of the most deprived wards in the country but the headteacher emphasised that only 3% of the parents were educated post-16 (and this was also reflected in the LSAs who the school was able to employ who were largely local mothers). The settlement, which is an inner suburb of the Northern city, was built as a coal mining settlement. The aforementioned quote highlights both the kinds of resource constraints and cultural elements of the local area, whilst at the same time providing some insights into the divergence of cultural expectations and assumptions of the 'middle-class' teachers compared to the socially excluded/working-class families. There is an interesting assumption of a commonality of viewpoints between the teachers with myself. Some of the comments made align to my own observations. I was warned by friends and family that this was a challenging and challenged part of the city, a city in which I was a Southern stranger. I fell in love with the children and the school context, where I was made to feel most welcome. I was also struck by the well-tended gardens and houses in this estate which was architecturally identical to the mining village in which I had grown up, designed by Abercrombie on garden village principles.

The strong kinship ties can act both to enhance feelings of social belonging but also to exclude those who do not belong. These kinships and social bonds can also lead to being 'stuck' in a particular place and can be accompanied by low educational and economic aspirations (Behtoui, 2017).

> The problem with the strong kinship networks is that it makes it difficult for people to get out. It's like Lina in year 6, she's a lovely kid and she's had a really hard life. The best thing she could do is leave this area and cut all ties. I always say to her, I want you to come and visit me one day and tell me you're at college. But she won't go to university. How can our kids ever hope to go to university? They just wouldn't be able to afford it! Not in a million years.
>
> (Ms. Trim[12])

It is interesting to note that the headteacher reinforced the low aspirations. Social mobility from these areas often must be accompanied by spatial mobility; however, they were also places with a great deal of community strength and good quality houses with gardens which are, by design, quality housing for the working masses with access to nature and community facilities. These communities had been subject to a rapid and revanchist (Neil Smith, 2005) decline in their traditional industries, and lives and livelihoods were (and continue to be) economically challenging and precarious.

The young people themselves discussed some of the challenges faced by their families, and how hard their parents worked, in low-paid, precarious employment. For instance, Loretta[13] discussed how her mum worked two jobs and her dad had a job in a tannery where he worked long hours, including nights:

> Yeah. My mum works at Saver Centre. Well, she's got two jobs actually. She works at the shop on her days off and um, she works at the Saver Centre. And my dad, he works at – you know where you take the skin off animals – leather skin, he does that. He works for a leather company. He works nights as well.

Many of the young people experienced family conflict and problems, such as parents' relationships breaking down, alcohol abuse, violence, and even abuse and neglect. There were some young people who were being monitored by social services for suspected neglect and one child, Rosie, had been removed from her family whilst at the school, as she had been abused and neglected; she was being fostered by a local family. Graham[14] talked about some of the issues his family had faced, and how this had forced his mum and his family to live in a location where vandalism and anti-social behaviour were pervasive:

> Cause my dad kicked us out. Cause one night, it were the day before Christmas, and my mum were on internet, and my dad came in and he were blabbering on about my mum seeing someone on the internet, and it were just one of my mum's mates that we were talking to, so we live in [a different area of the suburb] now. I still go and see my dad, so I go on a Friday. I stay on Friday night, and I stay on Saturday, but I don't stay on Saturday night, I stay at my grandmas. . . . Yeah, but we had this cage in our bedroom, so there's saw-dust all over. So we're gonna clean that up, and then we'll get a new carpet, and we have to decorate before we put the carpet down.

Graham went on to discuss how his family had moved to a different, nearby suburb, where the houses were bigger and there were issues with vandalism: "The houses are bigger, and there's people always smashing windows, but we haven't had ours smashed yet. Because – we went there, just before Christmas".

Similarly, Nicola[15] discusses some of the problems she has faced in her home life:

> Yep. My dad gets drunk, so my mum kicked him out, and I don't know where he is now, but he's in a flat or something. . . . Yeah. A good thing. Because on Lukie's birthday they got back together again, and we all celebrated a bit, and we all celebrated a bit, and my dad just went out of the house, and he went out of the house, and when he got back, Lukie was battering him and that, because he don't like my dad, Lukie doesn't. Lukie knocked his head off of the wall. [My dad] gets drunk and nasty. Yeah. When, I go home, it's all loud and I don't like it. And, when I'm trying to get to sleep, and I can't because it's loud.

Teachers and other staff in the school discussed the impacts that this broader context had on children, and one teaching assistant emphasised:

> "There are lots of children in year 6 with serious social problems. They might steal from the teachers . . . some of them have told me to restrain him [a boy walks out of school], but I won't – because I could easily break his bones by accident, because of my Taekwondo" (she went on to say) ". . . some of them restrain them – it happens a lot in year 6, but I won't do it".
>
> (Ms. Jessop[16])

Of course, the norms of the school and the everyday symbolic violence and assumption of the superiority of middle-class ways of being and knowing (Bourdieu, 1984) which led to commonplace and everyday use of actual violence against children were less critically examined than the issues the families faced. In addition, despite the fact that there may be a relationship between serious addiction, poverty and family problems (and this is not a linear relationship; the stress of poverty can exacerbate mental health and addiction issues), it is also true that poor families are much more subject to surveillance and scrutiny than middle-class families, who have more space in their homes and more places to hide their familial problems and any addictions (Featherstone et al., 2014, 2018; Lister, 2021; Cottam, 2018; Cross, 2021; Lens, 2019).

In general, teachers and other professionals were sympathetic to the situation of the families, whilst still othering them. Nonetheless, they were careful not to blame the families for their poverty or their difficulties. Blaming the poor for their poverty is an increasingly pervasive discourse (Shildrick and MacDonald, 2013), with poor people being cast as failures within the context of the brave new world of the risk society. In the risk society (Beck, 1992), traditional roles and expectations are eroded (e.g., the expectation that children would follow the life course and professions of their parents), opening up more apparent potential for social mobility and flexibility. Those that are not able to thrive in the risk society have their failures personalised. This denies the entrenched structural conditions which enable and constrain opportunity and social mobility, some of which are clearly present within the context of Church Street.

The adults in the school emphasised that although the parents loved their children and wanted to do the best for them, stresses tied to family life and poverty, alongside their own lack of education, meant that they were less able to support their children's education than more educated and affluent parents:

> Yes, we've got parents that . . . there are an awful lot of single parents here, and they have to try and work, because the government says they have to try and work. So, if you're getting Jimmy off to . . . to school in the morning, then going to work all day, coming home, and you've got to do all the things that . . . we do, cooking, cleaning and all the rest of it. You're tired, you plonk yourself . . . your kid in front of the video, you haven't got time to sit and read, and talk to the child. So the child gets no adult interaction, and you end up with, reduced language.
>
> (Ms. Trim[17])

8.3.2 Funding and resource constraints

My research diary recalls that the buildings were 'rundown and dilapidated'. One of the teachers explained what the school is:

> We are bottom of the pile [with] falling rolls and revenue. We lost a teacher and two LSAs this year and had to double up on class three and four.
>
> (Ms. Buttery[18])

Staff discussed the issue of the surplus of primary school places in the city. A new school with excellent new facilities had opened close to Church Street and many children were going there instead of Church Street. Since funding is provided per capita, any loss in children registered at the school is also a loss of income.

The resource constraints were exacerbated by the fact that the school had a high number of students experiencing SEND, but no Statements with the associated extra resources and powers (Ruth Lupton et al., 2010; Hutchinson, 2021). The first step in identifying a child as experiencing SEND is that they fall below their peers. The staff pointed to the fact that there was a low normative expectation on learning in the class, and therefore children who were falling well below national averages and expectations were not put forward for assessment for a Statement of SEND, as teachers emphasised:

> No, I think it was very different. If you compared children that are very well off to – children that we have, here, that we don't even register, would be on like – four or five in one of those schools, but because they're just normal here to be like that – so we only would Statement the absolute absolutes.
>
> (Ms. Gregson[19])

> people like Charlie and Jamie, who are eight years old and don't know their initial letter sounds. They can't read and they can't write, because it's kids like them and they don't even attract any money at all. It's bizarre.
>
> (Mr Keegan[20])

These resource constraints were contextualised within the marketisation of the LA's resources, in which the school had to purchase services from the LA, including support for human resources, educational psychology assessments and so on. This approach is now standard across the country. The headteacher emphasised the issue of high staffing costs tied to sickness leave which resulted from the stress of working in such a pressurised environment. Sickness absence was pervasive and expensive, as both the sick pay of the absent staff member (above statutory levels) and the replacement teaching cover came from the school budget. Some teachers claimed that the school leadership team were performing and managing badly. The limited resources had visceral impacts upon the young people's education. For instance, the following excerpt is from a Maths class in which there were 43 children:

> Then it was Maths with the middle set. **There were about 43 children in that class, and no additional support.** Although the planning was quite good, and differentiated, it was difficult to get round all the children, and one child just hadn't done anything. I helped out as an LSA. I asked whether it was usual for there to be so many children in the class, and the teacher told me that this was not unusual, and that class always has over 40 children.
>
> (Research diary, Church Street)

8.3.3 A normative, instrumentalist education institution versus alternative 'enriching' education

The headteacher was critical of the instrumentalist view of education emerging from the national education institution and government policy directives, measurable by SATs and published in league tables. She argued that the focus on employability was both narrow and irrelevant for many students in her school:

> There's more to life than just getting a job, and. . some of our children will never get jobs because they are . . . the children of third generation unemployed people. If you're giving them something that will enhance their life . . . if only gardening, so that they can enjoy planting their own seeds and growing their own vegetables, or flowers, or whatever, that is putting something into their lives, even if they don't get a job. Obviously, I want them to get jobs, but it seems to me that government policy is . . . you put in X hours of numeracy, you put in X hours of literacy, and you end up with a child who's got a level three or a level four Maths or English at the end of their primary school, and they will go on and get . . . X number of GCSEs, some of our children won't . . . you know, you can't expect

> a child with Down's syndrome to end up with five, grade one GCSEs, it's just not going to happen. But, if those children have had a wonderful time at school, and . . . enhanced the learning and the life experiences of the children around them, it's breaking down prejudice, it's doing all sorts of things, and that to me, is as valid – I'm not saying more valid, I'm saying as valid, and there should be room in education for both, not just the academic side, we should be looking at, the whole child. But that's a sixties teacher speaking.
>
> (Ms. Trim[21])

Her view demonstrates an appreciation of the wider potentials of education in broadening horizons beyond limited employment-focused skills. It also demonstrates and reproduces low educational and life course aspirations for the students in the school. This perception from the headteacher also ties into one of my enduring observations within schools: that ways of learning in schools often conflict entirely with research about how best to teach young people. This is not because teachers are not experienced and highly trained professionals who know how to inspire their students, rather it is that the content-heavy curriculum and testing regimes pressurise educators to teach to tests and cram content into every lesson.

8.3.4 Internal versus external impacts on low formal academic results and financial issues in the school – competing discourses

Although the headteacher of Church Street highlighted the contextual factors in the difficulties the school was facing, and the low SATs results, she also claimed:

> But, in the last two years they've had poor teachers, identified by Ofsted as poor teachers, and we've failed them. In previous years where they've had no supply teachers, where they've had good teachers right the way through, we've been within ten percent of the national average, and with the early years unit coming on stream, there's no reason why we shouldn't reach the national average. We've hit the national averages at key stage one this year, in fact, slightly above. There's nothing wrong with our children.

Again, however, Ms. Trim emphasised that the poor teachers themselves within the school were largely outside of her control. She went on to state:

> It can take six years to lose a teacher if you go down the incompetency root. Forget what the government says about fast-track dismissal, it doesn't happen in practice. Um . . . you have to put in support, you've got to do it for a minimum of two years, and that's two years, and that's two cohorts of children in their one crack at that particular year.

Other members of staff who still worked at the school, and a teacher who had previously worked in Church Street (who at that time was employed at Rose Hill), emphasised poor financial and educational management. Mr Taylor,[22] who now worked at Rose Hill, compared the two schools and pointed out:

> As far as the school itself, as far as I'm concerned, the organisation and the management, and the things we do is much more – successful as well, which comes across to the children. There's probably a much more coherent whole school approach to everything. I think there's a good quality of people in different posts, really. There's . . . vision at the top and then there's people that are in key posts, different co-ordinators and so on who help taking a lead in things and make sure that progress is maintained really. And they . . . provide good ideas that people would want to . . . to follow. I suppose sometimes the difficulty is that there are just so many things, that it is just isolating what's the thing to work on. A lot of the initiatives are government driven, but we like to pick up on them and do them the best we can, which obviously means there is a lot of work, you know. But I think generally people tend to pull together to make sure things are . . . are done. It's not competitive, it's usually collaborative.

This competing discourse about school effectiveness follows a broader tendency of external agencies to lay the blame on internal factors of school effectiveness, whereas internal managers emphasise resource constraints and human resources issues (Strand, 2016).

8.4 Rose Hill – an effective, mixed-intake school with a resourced facility for children with physical impairments

Rose Hill was selected as a case-study school through an examination of LA literature and discussions with LA actors. The school was selected as it had a high proportion of disabled children and a relatively high proportion of children with Statements of SEND. As a special resource school for physically disabled children, Rose Hill was within the LA's approach to 'inclusion'.[23] At the time of the study, eight children with physical impairments attended the school. Rose Hill, with 277 pupils, was a large primary school. The school had a mixed intake of pupils in terms of ethnic origin and socio-economic background; although fewer of the students came from socio-economically deprived backgrounds than Church Street, a significantly higher than average proportion of students were eligible for free school meals, which is considered to be an accurate, though imperfect, measure of poverty (Ilie et al., 2017). Like Church Street, Rose Hill was situated in an inner suburb of the city; although the area had a more mixed socio-economic demographic, the housing stock was similar. Rose Hill was a 'resource school' for physically disabled children and was generally a well-resourced school. A look at the school's current website demonstrates that it continues to be a well-resourced school with improved provision for young people with physical impairments.

The school was considered to be effective, achieving above-average levels in SATs. Although mixed, there were some children from particularly challenging backgrounds, and teachers discussed concerns that a particular boy, who also had labels of SEND, was being neglected. Social services had been involved, but the SENCO stated:

> Social services have been involved and everything, in making sure that he is properly looked after . . ., but there's nothing that can be proved really, and there's not been any evidence to show that he's not, but . . . [t]he school have clothed him basically, and . . . found him things and, this week we had trousers that were urinated on before Christmas that hadn't been washed and been put back on [after the Christmas vacation] . . . and um, it's very difficult.
>
> (Ms Robinson[24])

It was notable that the pace of learning and the quantity of the content children learned within a day was much greater in Rose Hill than Church Street. Teachers did emphasise that it was a good and supportive place to work, with an effective management structure; however, they also felt a constant pressure to be excellent, which could be challenging. The school was facing some contextual issues tied to high mobility of children and the financial insecurity tied to declining rolls across the city. The mixed intake of children included some with highly educated, often international families who attended the university as students, researchers and academic staff; however, these families were relatively mobile and their children often did not stay for the full seven years of primary school. The lack of financial stability had led to an inability to commit to new permanent posts, and all staff mentioned that there had been an over-reliance on temporary staff. Despite this challenging situation, the headteacher and other teachers in the school emphasised the role of the internal management, organisation and leadership of the school in setting high standards. As Mr Parker[25] stated:

> It is a very good school. We have commitment from the staff, focus from the staff, we focus on progress, and we've got good systems in place that support that. I think there's a very good working ethos in the school. In terms of monitoring progress, that's a very high thing in the school, there's also a lot of support in the school. I don't know if . . . those are the things that make a difference as well. I think it's the way that we whip them around. I mean you would have made some judgements as well. What would you say? It's the systems we have in place, the positive behaviour, the re-enforcing, the models the systems for good behaviour and good work. It's also the other – making sure that the children know that there are rules and they must obey them, and being consistent in that. And we do have those systems in place. Reinforcing the things that children do well and also picking up on things that they don't do well. But you always give children an opportunity. If they behave badly, but then they improve, they do get the opportunity. So yes, that work is important. Work is important because it determines where you are

> going in the future. So you have to keep saying that to children, have fun, work hard, it's very simple isn't it?

The individual agency and effectiveness of staff and leaders in the school was also accredited for the 'resource' status of the school as a specialist mainstream facility for children with physical impairments. This arrangement with the LA had come about ad hoc and was credited as being tied to the individual beliefs of the headteacher and his team and the ethos of the school, as Mr Parker emphasises:

> The other thing, to be honest Louise, is that it was never set up and it never has been. And what you find is, I mean for instance take, when I first set it up again, it had no toilets, no proper . . . so in other words we had one child who had difficulty with toileting. Fine when he was little, people could lift him on and off. As he got older, he grew long and gangly and dangly and impossible to move. So we had no toilet that was appropriate. We then had another little girl come in, no toilet that was appropriate. Well, I said, what do I do? Well, we looked for space, we were short of space. So we took under the stairs and the other end, which is the same as where Barbara works now, our caretaker, if you look at her cupboard here, that's what the other end was like. You have a look at it, because all I did was to say "look, we could block that in and turn it into a toilet. Not suitable I was told." So I said we've got nowhere else, so we had better try it. Now go and have a look, because . . . So what did we have? We had a ramp, we had one ramp into school. That wasn't put there until we asked for it. We then had a toilet, which was after a few years. But now when you look, what have we got? We've got proper toilets, proper facilities, a hoist . . . Physio room, now that came later. We suddenly found, you see, that children were doing physio in the staffroom. The staffroom was busy, where could they find a space? It was undignified. So we looked for a room, what did we do? We turned a cloakroom into a physio room. So we built an ITC suite, couldn't get the children in with wheelchairs. What did we do? We looked to extend it . . . so it was never planned. It's a living sort of way of going on really, as we find a need, then we try and meet that and address that need. And that's the reality. I mean the idea of it being set up just isn't true. It's like the field, you know, we just put a ramp down. That was done last summer, not, not eight years ago, you know, so it's sort of yes, lift first, was the first thing we had in terms of – physical resources. The next thing was the toilet and it's come on from there really. And the next thing was – well we've got one ramp, how do they get into this area, how do they get into that area, so we had other ramps built. So it's a sort of evolving thing really . . . the LA have a great model here which could be used for other areas, but it hasn't really been taken up. But this works.

This quote is one example of why the critical question about how to educate young people with SEND is not necessarily about a dichotomy between special and mainstream schools, but how these two institutions intersect and connect to most effectively support young people with SEND labels and/or mind-body-emotional differences.

8.5 Poverty, inequality, limited social capital and disability: wasted lives

In the accounts from Church Street above, and indeed across many of the schools which were the focus of the research across all the studies, there was an implicit or explicit connection made between poverty, socio-economic hardship, low levels of parental education/cultural capital and SEND. This is specifically understood to be the case in relation to learning differences, and particularly non-specific labels of learning difference (see also Holt et al., 2019a). These discussions reflect broader academic debates, which demonstrate disproportionality in experiences and labels of SEND along intersecting class, gender (more boys are labelled with/experience SEND, although girls' experiences are therefore specifically gendered) and racial/ethnic grounds (Banks et al., 2012; Dyson and Gallannaugh, 2008; Youdell, 2010; Cruz and Rodl, 2018; see also Holt et al., 2019a and Azpitarte and Holt, 2023). In relation to learning differences, a (white, male) educational psychologist from a city in the Southeast of England summed up a widely held belief:

> There is an argument to be made that there are actually some learning difficulties which are more-poverty related than anything else. Moderate learning difficulties and behaviour chiefly. And actually I did a mapping exercise . . . part of my study was mapping of disabilities and moderate learning difficulties . . . when you analysed them, more than 95% of them came from areas associated with hardship, more than 95%. . . . I went back to definitions of moderate learning difficulties because I was just so shocked – it seemed to me that actually an economist could make a better prediction about MLD than an educational psychologist.

As the educational psychologist cited earlier emphasised, moderate learning differences are also closely associated with poverty. In a previous paper with Sophie Bowlby and Jennifer Lea (Holt et al., 2019) we identified how class and capitals intersect with SEND, such that those who are more educated and have more resources fight to gain a specific label of SEND for their child (see also Riddell and Weedon, 2016). Therefore, it is highly likely that the label of moderate learning disabilities could conceal a range of specific learning differences or neurodiversities, such as dyslexia, dyscalculia, motor coordination problems and so on, which, if properly supported and with reasonable adjustments in place, would be no impediment to learning. This possibility is further suggested by the research I have recently undertaken with my colleague Fran Azpitarte (Azpitarte and Holt, 2023). In this paper we map and model the educational outcomes of young people with labels of SEND at age seven, entirely within the context of the new Code of Practice. We find that the outcomes are both shockingly low and geographically variable by Local Authority. We also find that the poorest performing Local Authorities have the highest proportion of young people with labels of moderate learning differences, suggesting further that this category actually represents an amorphous group of young people whose needs, and support requirements have not been appropriately identified or supported. In that context, it is perhaps unsurprising that such young people are not achieving well at school. Additionally, the increasingly instrumental, recording and content-focused nature of the curriculum is such that it supports the thriving of a very small proportion of young people.

In addition to this disproportionality, more young people with impairments lived in poor families in the research, and this also reflects broader research (Bradshaw and Main, 2016) and official statistics (Joseph Rowntree Foundation, 2022). There might be a variety of interconnected reasons for this pattern; however, one factor is likely to be tied to the fact that there are high direct and indirect costs to families with disabled children. Directly, parents (especially mothers) might have significant caring duties which limit and constrain their work in paid employment, and disabled children have specific resource requirements, such as therapies, adaptations and specific equipment. Benefits are inadequate and decreasing (Saffer et al., 2018; Blackwell, 2022). The pressures of Austerity have made this disproportionality, and the real hardships faced by families with disabled children, even starker. For disabled people, this has become what I will label a politics of genocide, as disabled people's lives are rendered unliveable in a context where neoliberal governments retrench state benefits of the non-working benefits claimants – who are overwhelmingly disabled people and those with mental ill-health (Ryan, 2020; see also Edwards and Maxwell, 2023). Currently within the UK, as in many contexts of the globalised world, disabled people are not being given the contexts for 'liveable lives', which Butler suggests would include:

> having proper support, and that includes the economic condition of persisting in life, and in reproducing the material conditions of life. Shelter, food, employment all count here. At the same time, certain kinds of freedoms, such as assembly, mobility, and expression, are also part of liveability.
>
> (Judith Butler in Zaharijević, 2016:111)

Indeed, the devaluing of disabled people's bodies and lives has parallels to the invisibility of the deaths of people across the majority world, which Butler (2004b) discusses. Disabled people are rendered increasingly invisible and isolated (Dodd, 2016; Burch, 2018). The topic of disability is also sidelined within political debates in which their benefits and claims on society are eroded as they are labelled as 'non-working benefits claimants', even though disabled people make up the majority of this category.

Within this context, decisions about whether to keep or abort a foetus with impairments are coloured by the prevailing disablist discourses, and the stark reality of the visceral horrors of disabled people's lives if they do not have means of support beyond that provided by the state. This is a picture which plays itself out in schools, where a lack of real inclusion and appropriate therapies renders meaningful attainment for children with SEND almost impossible for families without the capital to invest in private education and therapies, or to battle to get their children the support they require (Blackwell, 2022).

8.6 Reflections

In this chapter I have provided some context regarding the ways that young people's social relationships are differentially positioned in relation to broader axes of power and access to resources and capitals. I have reflected on how the spaces in schools, and within which schools are positioned, can constrain and enable young people's social relationships, and the extent to which these relationships provide access to capitals – social, cultural and economic. Where schools are – the areas in which they are located, the way they are positioned within wider institutional frameworks and resources of education, social support and health, and who they recruit as teachers, teaching assistants and leaders – is critical. This intersects with the young people's own subjectivities, histories and trajectories. Some young people and their schools and neighbourhoods are clearly disinvested, and cast as "waste" (Katz, 2011, 2018). Certainly, the opportunities offered by their schooling and their social relationships do not always secure them a stake in the neoliberal risk society, and the young people's social relationships offer various access to broader capitals – cultural, economic and social. The young people in Church Street, even if they were included in school, were largely friends with other young people from poor and socially excluded backgrounds. They were relatively unlikely to be achieving age-related expectations in learning at the end of primary school. University was a distant dream. Nonetheless, all of the young people, their families and the educational professionals around them, had lively agencies which carved out alternative positive frameworks of being in the world, despite some of the substantial challenges they faced.

Notes

1 Lindsay, white British working-class girl with physical impairment who used a wheelchair, Rose Hill, year five.
2 Noel, white British boy from a poor family with many challenges and social services involvement, who often came to school dirty, and with non-specific learning differences and some motor coordination differences, Church Street, year five.
3 Ms. Miller, white British working-class LSA, Church Street.
4 Nelson, white working-class boy with learning and communication differences, Rose Hill, year five.
5 Mr Taylor, white British male class teacher, Rose Hill
6 Mr Parker, headteacher, white British male, Rose Hill.
7 Ms. Gregson, SENDCo, white British female, Church Street
8 All white British female teachers, Church Street.
9 Ms. Trim, headteacher, white British female, Church Street.
10 White British female senior teachers, Church Street.
11 Ms. Massey, white British female class teacher, Church Street.
12 Ms. Trim, headteacher, white British female, Church Street.
13 Loretta, white British working-class girl on the Autism Spectrum, Church Street, year five.

14 Graham, white British boy, from a poor/socially excluded background, with specific learning differences, Church Street, year five.
15 Nicola, white girl from a Traveller background who now lived in a house, with some learning differences but no label, Church Street, year four.
16 Ms. Jessop, white British female LSA, Church Street
17 Ms. Trim, headteacher, white British female Church Street.
18 Ms. Buttery, white British female teacher Church Street
19 Ms. Gregson, SENCO, white British female, Church Street.
20 Mr Keegan, white British male class teacher, Church Street.
21 Ms. Trim, headteacher, white British female, Church Street.
22 Mr Taylor, white British male class teacher, Rose Hill.
23 Based around impairment specific 'resource schools' and partnerships between special and mainstream schools.
24 M*s* Robinson, white British female class teacher, Rose Hill.
25 Mr Parker headteacher, white British male, Rose Hill.

9 Conclusions

This chapter draws together the major contributions of the book: the importance of young people's own socialities to their experiences of school, and how young people are critical in the reproduction and potential transformation of enduring axes of difference and entrenched disadvantages around class, dis/ability, race/ethnicity, gender and other differences, which intersect. The way that the original and new concepts *contextual bodies/subjectivities/agencies* and *immersive geographies* instigate new directions for social sciences and education is considered. As *contextual bodies/subjectivities/agencies*, young people are embodied yet contextual and dynamic in nature: young people *become* differently in different social, spatial, historical, political, economic and cultural contexts in interaction with their corporeality. As *nodes of the intergenerational reproduction of enduring differences*, young people frequently reproduce axes of power relations, which precede them and continue through time and space, and through which entrenched intergenerational differences continue. Yet finally, the book emphasises the *powerful nature of young people* and specifically their socialities and their power to reproduce but also *to challenge and change enduring broader-scale inequalities via their everyday performances*. The concepts of embodied emotional and social capital express this powerful and contextual nature of young people's subjectivities and emphasises how they are forged within (and can challenge and change) broader socio-spatial contexts. The concept of *immersive geographies* provides original perspectives about how repeated proximity through space and time provides opportunities for young people to challenge and change enduring axes of power, which has implications much beyond the scope of this book. The book is a countertopographical or empowering project, and by reflecting upon the conditions of the emergence of socio-spatial relations which defy enduring embodied inequalities, some ways that schools can be empowering to young people are suggested. The book concludes with my original poem: A Circle.

The book has foregrounded young people's social experiences in schools and has demonstrated how pivotal and central young people's social experiences are in schools and to social reproduction and, potentially, transformation. As the quotes that preface the book emphasise, society invests in young people a futurity and hope that if we put together young people who are 'different' on the grounds of a host of intersecting axes of power relations, often race and or dis/ability, but

DOI: 10.4324/9781003028161-9

also class, gender, sex and sexuality and so on, these divisions will be played out differently in the future. However much we might admire Rev Dr Martin Luther King or Dame Louise Casey, it is clear that this is an overly idealised notion which fails to recognise young people's own contextual bodies/subjectivities/agencies. Yet, young people's social relationships do have the power to transform, and often more frequently, reproduce enduring entrenched embodied social differences.

The book has emphasised how important young people's own social relationships are to the conundrum that education provides social opportunities to enhance wealth and the future trajectories of young people and yet so often they (also) reproduce enduring inequalities tied to class, wealth and poverty-based disadvantage and advantage, dis/ability, race/ethnicity and gender. Young people's social relationships are critical and potentially transformational in the ways in which powerful categories, which frame a host of intersecting dis/advantages and inequalities in societies, are understood and performed relationally and can, potentially be challenged and changed. Young people's social capital and cultural capital intersect, as in the main, friendships were critically important to engagement in school; those young people who were excluded, isolated or marginalised were often also those who had more negative experiences with school more broadly. This relationship was not one-directional, and those who had more difficult experiences with formal aspects of school or with teachers and other adults often also had difficult relationships with peers, although often on different grounds.

These central conceptual insights are demonstrated through examining young people's friendships, which provide embodied emotional and social capital. In Chapter 5, drawing upon young people's accounts, I demonstrated how important young people's friendships are to them, and how they provide emotional reciprocity, connection and affirmation. It seems that frequently young people's friendships operate within a dynamic tension between connection and disassociation and wanting to have close social bonds and wanting to, or fearing that friends, make connections with others. Some friendships were diffuse and less intense, and there seemed to often be a gendered element to how intense friendships were *reported to be*, although the same boys who claimed to be not too concerned about who they played with consistently photographed each other and discussed the same close friends. The importance of friendships to the majority of young people provided a context of 'recognition' enabling the creative play of power which is subjection.

Taking this theme forward, Chapter 6 reflected on young people as nodes on the intergenerational reproduction of enduring differences. Reviewing the interviews, the ethnographic research and some of the more creative methods and artefacts of young people, I reflected on some of the patterns of young people's social relationships which forge particular subjectivities. These move from some subtle power plays, such as the roles taken in different games which have racial or ableist undertones, to very active processes of exclusion and isolation. From the analysis it became evident that for many young people, friendships were foundational to their experience of school, and in particular those who were, or had previously been, excluded and isolated also had negative experiences of school more generally in the formal of curricula and/or negative relationships with teachers, who can also exclude, stigmatise and marginalise young people. Sometimes this was clearly conscious and deliberate. More often, this was through subtle normative acts tied to expectations of behaviour or learning with which some young people could not comply. It is important to note, however, that Jennifer in particular was struck that we should not normalise the importance of friendships to all young people; indeed, for some young people school was all and only about the curricula.

The original idea of ***immersive geographies*** explores the ways in which the coming together of (young) people through repeated encounters in space and time provides a potential for of the ways in which difference is performed, enacted and understood. This difference can be tied to enduring, and intersecting, axes of power, such as class, race/ethnicity, religion, gender/sex, sexuality and my specific research interest of dis/ability, alongside more subtle differentiations. School spaces are specific, as young people come together repeatedly and enduringly. The space/time/space/time/space/time dialectic of repeated encounters in spaces which are ostensibly the same and yet performed slightly differently every time is critical to immersive geographies. The bringing together of young people in specific spaces repeatedly allows them to forge deep and affective connections which transform them in some way. This sense of new open possibilities and the ability to generate new worlds, which parallels immersive virtual technologies, is critical to immersive geographies. In schools (young) people converge to do similar things day after day, and yet every time they come together the connection and the practice is a performance; it is provisional, and new realities are made every day. Although I critique the simplistic assumption that co-locating young people will automatically transform relationships between groups who have been divided and occupy hierarchical relationships to each other, I also argue that there is an imminent potential to transform these enduring axes of power inherent in the provisionality of every performance between young people as they encounter each other in space through time. These specific, yet repeated, circular moments of time taking place day after day with the same people in the same place doing similar, yet subtly different, things, are never fully left behind when the young people leave the specific, dynamic, space/time of the school. They become an embodied part of young people's subjectivities, which is then sedimented within the memories and habitual ways of being of the young people. Young people are forged in these spaces in ways which are, perhaps, never fully left behind and in some way are formative to their subjectivities. The space/time/space/time/space/time re-enacted, the everyday and repeated coming together of people and things in the same space, performed in similar and yet often subtly or more profoundly different ways, is never isolated to that space/time. It connects to space/times before and after, and to space/times beyond the porous boundary of schools and a host of intersecting scales, from the local to the global.

These immersive geographies are open, porous and connected. They happen in specific places (Massey, 1993, 2005), which are situated within constellations of powers and resources, which constrain (although do not determine) the potentials of the young people's powers. In Chapter 8, I trace some of the geographical contexts of the young people's lives, and this points to the ways in which their lives are constrained and enabled by their positioning within broader relations of power and resources. I hope this is not reductive, and yet I am struck by how differently the young people are positioned and the ways in which political and economic processes are framing and constraining young people's potentials. Their social and emotional capitals provide very different access to opportunities and resources. This chapter is inspired by the more Marxist positions emerging from some North American scholars of young people (notably Katz, e.g., 2004). In addition, this focus has been driven by the stark realities of the revanchist capitalism, which I have been forced to confront by the gathered stories of these young people and the news I read and listen to every day. In Chapter 8, I explore the socio-spatial contexts of two primary schools within the same city. I then go on to tease out some 'contour lines' that connect disparate places similarly constituted or affected by certain problems' (Katz, 2008: 25). In particular, I focus on the intersections between poverty, inequality, limited social capital and disability, arguing that through systematic processes of disinvestment and (ab)normalisation there is an intergenerational 'waste' of young people who are being failed by the education system because they are poor and have mind-body-emotional differences.

More hopefully, this very connectedness provides opportunities for some of the more empowering ways in which subjectivities are played out to have an impact beyond the immediate space/time of it happening. I have two key ideas of how this is so. First, the young people as ***contextual bodies/subjectivities/agencies*** have trajectories beyond the school as they move into other spaces and times. We can hope that these new radical ways of being together become part of the young people's habitus, a beyond-conscious backdrop to their future social engagements. So, as the young people move through space and time, they are taking these experiences with them in conscious or beyond-conscious ways. In the introduction, I reflected upon my trajectory, and similarly young people's memories and experiences become an embodied part of their subjectivities. Our young selves are never quite left behind.

Second, the book is, drawing again upon Katz (e.g., 2001, 2004) a **counter-topographical project**. The book demonstrates how young people in different schools experience similar processes of subjection around powerful and enduring axes of power. More critically, the book foregrounds moments in which young people exceed and challenge the exegesis of power, and become something else, radically decentring regimes of dis/ableism, class power, sex/gender, and racial and ethnic inequalities. For the most part, these ways of being are moments in space and time unconsciously performed and were just about playing – playing something completely separate and independent from 'real' life or playing with subjectivities and ways of being in the world. The immersive moments with radical potentials to be otherwise were largely not self-reflected upon by the young people, such as the child-wheelchair-child-child-child-child-become horsie. They have been taken away and analysed by me and others. I have also reflected upon the conditions of the emergence of these moments in which new ways of being emerge, to reflect upon whether these can be recreated in other space/times. As such I hope that some of these moments that exceed the exegesis of power that constrain powerful subjectivities, and the conditions for that exceeding, inspire similar ways of being in other contexts beyond those very specific sites in which they were played out.

There are some examples in the book of how to encourage more positive connections between young people categorised as 'different' in some way to each other, and it goes beyond just putting them in the same physical space. I want to take the opportunity to sum up what schools and others can do to enhance the opportunities for new lines of flight. Foreground young people's socialities; give young people space to be convivial around shared interests. Be aware that some young people are vulnerable to bullying, exclusion and marginalisation, because of their mind-body-emotional characteristics and labels attached to them, because they come from poor or socially excluded families, they might smell, because laundry is expensive and families are choosing between food, heating, water and might not have access to a washing machine. Challenge racism and be aware of our own conscious and subconscious biases. Be enabling rather than ableist and think about how you can challenge the normally developing child. Prioritise the friendships of young people who might be vulnerable to exclusion, but do not try to force friendships. It is not natural or normal for children and young people to be isolated, bullied, left out or stigmatised, so I urge the reader to challenge these practices for all young people from all backgrounds with the whole range of mind-body-emotional characteristics. We should not punish young people for behaving in ways that are part of their mind-body-emotional corporeality by further excluding them from important spaces of sociality. Use restorative justice approaches instead. Finally, let us all resist the increasing pressure to become Gradgrinds and remember the lively affective learning possibilities that attracted us to work with children and young people and/or to be educators and scholars.

A Circle

Should I wrap this book up with a neat summary?
We met these children in a moment of their journey:
Where are you now?
Are you a soldier, a sailor, a candlestick maker,
Narcotic dependent, a mother, a father, a friend?
But, blue-eyed Nadia
You were full flight to the sun;
She lowered her gaze to you!
And your brilliance shines Lindsay.
Bright, quick full wick with life;
Yet, pain seared every movement, you couldn't race around.
You thought you did not fit in.
No, you exceeded! You sang with a voice from the gods.
What a tragedy; they said.
What unbridled joy.
And you Ben, with your soiled trousers
Left in your bag since before Christmas.

I hope you managed to escape
Your beginnings.
You were kind and soft and everyone's friend.
Rhana, Queen of the hijabed playground gang
Who pulled at my arms and swirled me around,
Whispered of futures in desert states,
Just here for a short stay with Professor mum or dad.
You kept to your group;
Were you pushed or pulled?
Sparkling bright in the classroom and playground.
Finally, to Ali, your nine-year-old self is trapped in my heart,
Racing around in your motorised cart.
With the friends trying to catch you,
Trailing in your wake.
Zooming through doors on full speed,
Whilst the other children wait.
For someone to open the heavy weight,
No, not you – you power through!
Yet, here's to the children left behind,
In crumbling schools,
At the back of the class,
In corridors,
In young offender's dining halls,
On the street,
In damp, dark homes,
That cost the world,
And yet without a stake
In this green and sceptred, unequal isle.
Did you achieve the ambition of becoming
a teaching assistant?
Did you reach those dizzy hights, Naomie.
I hope so and more.
Here is a prayer to the children-now-adults I met,
To the others I haven't yet.
Here's to the dream of the life to come.
Here's to play,
To fun.
To every time you move,
You speak,
You take a chance,
It is a world undone,
Done differently.
Done anew.
You go beyond the terrestrial plane,
To a flight of fancy.

Child-wheelchair-child-child-child-child assemblage
In your gravity-defying horsie game.
Was this one moment in time?
Did it happen every day?
Circular time, again and again.
But a line of flight to another plane of possibility.
Where are you now child-wheelchair-child-child-child-child become horsie?

Louise Holt

References

Adams, R. (2019). Children with special needs are marginalised at school, says NAO. *The Guardian*. www.theguardian.com/education/2019/sep/11/children-with-special-needs-are-marginalised-at-school-says-nao

Ahmed, S. (2013). *Strange Encounters: Embodied Others in Post-Coloniality*. London/New York: Routledge.

Aitken, S. C. (2001). *The Geographies of Young People: The Morally Contested Spaces of Identity*. London/New York: Routledge.

Aitken, S. C., & Herman, T. (1997). Gender, power and crib geography: Transitional spaces and potential places. *Gender, Place and Culture: A Journal of Feminist Geography*, *4*(1), 63–88.

Aitken, S. C., & Wingate, J. (1993). A preliminary study of the self-directed photography of middle-class, homeless, and mobility-impaired children. *The Professional Geographer*, *45*(1), 65–72.

Åkerblom, A., & Harju, A. (2021). The becoming of a Swedish preschool child? Migrant children and everyday nationalism. *Children's Geographies*, *19*(5), 514–525.

Alanen, L., Brooker, L., & Mayall, B. (2015). *Childhood with Bourdieu*. Amsterdam: Springer.

Alderson, P., & Morrow, V. (2011). *The Ethics of Research with Children and Young People: A Practical Handbook*. Thousand Oaks, CA: Sage.

Allan, A., & Charles, C. (2014). Cosmo girls: Configurations of class and femininity in elite educational settings. *British Journal of Sociology of Education*, *35*(3), 333–352.

Allen, J. (2011). *Lost Geographies of Power*. New York: Wiley.

Amin, A. (2002). Ethnicity and the multicultural city: Living with diversity. *Environment and Planning A*, *34*(6), 959–980.

Amin, A. (2006). The good city. *Urban Studies*, *43*(5–6), 1009–1023.

Anderson, B. (2006). Becoming and being hopeful: Towards a theory of affect. *Environment and Planning D: Society and Space*, *24*(5), 733–752.

Anderson, B. (2017). *Encountering affect: Capacities, apparatuses, conditions*. London/New York: Routledge.

Anderson, B., & Harrison, P. (2010). The promise of non-representational theories, in Anderson B., & Harrison, P. (eds) *Taking-place: Non-Representational Theories and Geography* (pp. 1–36). Farnham: Ashgate.

Anderson, C. S. (1982). The search for school climate: A review of the research. *Review of Educational Research*, *52*(3), 368–420.

Andrews, J., Robinson, D., & Hutchinson, J. (2017). *Closing the Gap?: Trends in Educational Attainment and Disadvantage*. London: Education Policy Institute.

Ansell, N. (2009). Childhood and the politics of scale: Descaling children's geographies? *Progress in Human Geography*, *33*(2), 190–209.

Arnold, M. (2011). *Thor: Myth to Marvel*. London: Bloomsbury Publishing.

Azpitarte, F., & Holt, L. (2023). Failing children with special educational needs and disabilities in England: New evidence of poor outcomes and a post-code lottery at the local authority level at key stage one. *British Educational Research Journal*. Online early https://doi.org/10.1002/berj.3930

Ball, S. J. (2003). *Class Strategies and The Education Market: The Middle Classes and Social Advantage*. London/New York: Routledge.

Ball, S. J. (2013). *The Education Debate*. Oxford: Policy Press.

Ball, S. J. (2018). The tragedy of state education in England: Reluctance, compromise and muddle-a system in disarray. *Journal of the British Academy*, *6*, 207–238.

Ball, S. J., & Vincent, C. (2007). Education, class fractions and the local rules of spatial relations. *Urban Studies*, *44*(7), 1175–1189.

Banks, J., Shevlin, M., & McCoy, S. (2012). Disproportionality in special education: Identifying children with emotional behavioural difficulties in Irish primary schools. *European Journal of Special Needs Education*, *27*(2), 219–235.

Beck, U. (1992). *Risk Society – Towards a New Modernity (Translated by Mark Ritter)*. Thousand Oaks, CA: Sage.

Behtoui, A. (2017). Social capital and the educational expectations of young people. *European Educational Research Journal*, *16*(4), 487–503.

Benabou, R., & Tirole, J. (2003). Intrinsic and extrinsic motivation. *The Review of Economic Studies*, *70*(3), 489–452.

Benjamin, J. (1998). *The Shadow of the Other: Intersubjectivity and Gender in Psychoanalysis*. London/New York: Routledge.

Benwell, M. C. (2014). From the banal to the blatant: Expressions of nationalism in secondary schools in Argentina and the Falkland Islands. *Geoforum*, *52*, 51–60.

Beresford, B. (1997). *Personal Accounts: Involving Disabled Children in Research*. York: Social Policy Research Unit.

Berkowitz, R., Moore, H., Astor, R. A., & Benbenishty, R. (2017). A research synthesis of the associations between socioeconomic background, inequality, school climate, and academic achievement. *Review of Educational Research*, *87*(2), 425–469.

Bessell, S. (2017). Rights-based research with children: Principles and practice, in Evans, R., Holt, L., & Skelton, T. (eds) *Methodological Approaches* (pp. 23–46). Berlin: Springer.

Bhana, D., & Mayeza, E. (2016). We don't play with gays, they're not real boys . . . they can't fight: Hegemonic masculinity and (homophobic) violence in the primary years of schooling. *International Journal of Educational Development*, *51*, 36–42.

Blackwell, C. (2022). The needs of children on the autism spectrum and their families: Exploring household costs and factors impacting access to resources (Doctoral dissertation, Loughborough: Loughborough University).

Blazek, M. (2021). Children's right to be hostile: Emotions and agency through psychodynamic lens. *Emotion, Space and Society*, *41*, 100850.

Bondi, L. (2002). Empathy and identification: Conceptual resources for Feminist fieldwork. *ACME: An International Journal for Critical Geographies*, *2*(1). www.acme-journal.org/index.php/acme/article/view/708

Bondi, L. (2005). Making connections and thinking through emotions: Between geography and psychotherapy. *Transactions of the Institute of British Geographers*, *30*(4), 433–448.

Bondi, L. (2014a). Understanding feelings: Engaging with unconscious communication and embodied knowledge. *Emotion, Space and Society*, *10*(1), 44–54. https://doi.org/10.1016/J.EMOSPA.2013.03.009

Bondi, L. (2014b). On Freud's geographies, in Pile, S., & Kingsbury, P. (eds) *Psychoanalytic Geographies* (pp. 57–72). London/New York: Routledge.

Bondi, L., Davidson, J., & Smith, M. (2007). Introduction: Geography's "emotional turn", in Davidson, J., Smith, M., & Bondi, L. (eds) *Emotional Geographies* (pp. 1–18). Farnham: Ashgate.

Bourdieu, P. (1977). *Outline of a Theory of Practice*. Cambridge: Cambridge University Press.

Bourdieu, P. (1984). *Distinction: A Social Critique of the Judgement of Taste*. Cambridge, MA: Harvard University Press.
Bourdieu, P. (1990). *The Logic of Practice*. Stanford: Stanford University Press.
Bourdieu, P. (2018). The forms of capital, in Granovetter, M., & Swedberg, R. (eds) *The Sociology of Economic Life* (pp. 78–92). London/New York: Routledge.
Bourdieu, P. (2020). *Habitus and Field: General Sociology, Volume 2 (1982–1983)*. Cambridge: Polity Press.
Bourdieu, P., & Passeron, J. C. (1979). *The Inheritors: French Students and their Relation to Culture*. Chicago: University of Chicago Press.
Bourdieu, P., & Passeron, J. C. (1990). *Reproduction in Education, Society and Culture*. Thousand Oaks, CA: Sage.
Bourdieu, P., & Thompson, J. (1991). *Language and Symbolic Power*. Cambridge, MA: Harvard University Press.
Bowlby, S., Lea, J., & Holt, L. (2014). Learning how to behave in school: A study of the experiences of children and young people with socio-emotional differences, in Mills, S. & Kraftl, P. (eds) *Informal Education, Childhood and Youth: Geographies, Histories, Practices* (pp. 124–139). London: Palgrave Macmillan UK.
Bradshaw, J., & Main, G. (2016). Child poverty and deprivation, in Bradshaw, J. (ed) *The Wellbeing of Children in the UK* (pp. 31–70). Thousand Oaks, CA: Sage.
Bridge, G. (2006). It's not just a question of taste: Gentrification, the neighbourhood, and cultural capital. *Environment and Planning A*, *38*(10), 1965–1978.
Brotherton, D. (2015). *Youth Street Gangs: A Critical Appraisal*. London/New York: Routledge.
Burch, L. (2018). 'You are a parasite on the productive classes': Online disablist hate speech in austere times. *Disability & Society*, *33*(3), 392–415.
Burgess, S., & Wilson, D. (2005). Ethnic segregation in England's schools. *Transactions of the Institute of British Geographers*, *30*(1), 20–36.
Butler, J. (1990). *Gender Trouble: Feminism and the Subversion of Identity*. New York: Theatre Arts Books.
Butler, J. (1993). *Bodies That Matter: On the Discursive Limits of "Sex."* London/New York: Routledge.
Butler, J. (1997). *The Psychic Life of Power: Theories in Subjection*. Stanford: Stanford University Press.
Butler, J. (1999). Performativity's social magic, in Shusterman, R. (ed) *Bourdieu: A Critical Reader* (pp. 113–128). Oxford: Blackwell.
Butler, J. (2004a). *Undoing Gender*. London/New York: Routledge.
Butler, J. (2004b). *Precarious Life: The Powers of Mourning and Violence*. New York: Verso.
Calarco, J. M. (2014). Coached for the classroom: Parents' cultural transmission and children's reproduction of educational inequalities. *American Sociological Review*, *79*(5), 1015–1037.
Callard, F. (2003). The taming of psychoanalysis in geography. *Social and Cultural Geography*, *4*(3), 295–312.
Cameron, D. (2015, July 20). *Extremism: Speech Delivered at Ninestiles School in Birmingham*. London: Prime Ministers' Office.
Canguilhem, G. (1973). *The Normal and the Pathological*. New York: Zone Books.
Carnes, N., & Goren, L. J. (2022). An introduction to the politics of the Marvel Cinematic Universe, in Carnes, N., & Goren, L. J. (eds) *The Politics of the Marvel Cinematic Universe* (pp. 1–18). Lawrence: University Press of Kansas.
Chatzitheochari, S., Parsons, S., & Platt, L. (2015). Doubly disadvantaged? Bullying experiences among disabled children and young people in England. *Sociology*, *50*(4), 695–713.
Children's Society (2022). *Special Educational Needs and Risk of School Exclusion*. London: Children's Society.
Chouinard, V., Hall, E., & Wilton, R. (eds) (2016). *Towards Enabling Geographies: "Disabled" Bodies and Minds in Society and Space*. London/New York: Routledge.

Christensen, P., & Prout, A. (2002). Working with ethical symmetry in social research with children. *Childhood*, *9*(4), 477–497.
Cockayne, D. G., Ruez, D., & Secor, A. J. (2020). Thinking space differently: Deleuze's Möbius topology for a theorisation of the encounter. *Transactions of the Institute of British Geographers*, *45*(1), 194–207.
Coleman, J. S. (1987). Social Capital and the development of youth. *Momentum*, *18*(4), 6–8.
Colls, R. (2012). Feminism, bodily difference and non-representational geographies. *Transactions of the Institute of British Geographers*, *37*(3), 430–445.
Connolly, P. (2002). *Racism, Gender Identities and Young Children: Social Relations in a Multi-Ethnic, Inner City Primary School*. London/New York: Routledge.
Cortés-Morales, S., Holt, L., Acevedo-Rincón, J., Aitken, S., Ladru, D. E., Joelsson, T., Kraftl, P., Murray, L., & Tebet, G. (2021). Children living in pandemic times: A geographical, transnational and situated view. *Children's Geographies*, *20*(4), 381–339.
Cottam, H. (2018). *Radical Help: How We Can Remake the Relationships Between Us and Revolutionise the Welfare State*. London: Virago.
Cross, T. L. (2021). Racial disproportionality and disparities among American Indian and Alaska native populations. *Child Maltreatment: Contemporary Issues in Research and Policy*, *11*, 99–124.
Cruz, R. A., & Rodl, J. E. (2018). An integrative synthesis of literature on disproportionality in special education. *The Journal of Special Education*, *52*(1), 50–63.
Das, R. J. (2004). Social capital and poverty of the wage-labour class: Problems with the social capital theory. *Transactions of the Institute of British Geographers*, *29*(1), 27–45.
Davidson, J., & Parr, H. (2014). Geographies of psychic life, in Pile, S., & Kingsbury, P. (eds) *Psychoanalytic Geographies* (pp. 119–134). London/New York: Routledge.
De Freitas, S., Rebolledo-Mendez, G., Liarokapis, F., Magoulas, G., & Poulovassilis, A. (2010). Learning as immersive experiences: Using the four-dimensional framework for designing and evaluating immersive learning experiences in a virtual world. *British Journal of Educational Technology*, *41*(1), 69–85.
Dear, M., Wilton, R., Gaber, S. L., & Takahashi, L. (1997). Seeing people differently: The sociospatial construction of disability. *Environment and Planning D: Society and Space*, *15*(4), 455–480.
Deleuze, G., & Guattari, F. (1988). *A Thousand Plateaus: Capitalism and Schizophrenia*. London: Bloomsbury Publishing.
Delgado, M. (2015). *Urban Youth and Photovoice: Visual Ethnography in Action*. Oxford: Oxford University Press.
Department for Education. (2007). *Statistics: Special Educational Needs*. HMSO: London.
Department for Education. (2015). *Statistics: Special Educational Needs*. HMSO: London.
Department for Education. (2022). *Integrated Schools*. HMSO: London.
Dickens, C. (1854). *Hard Times*. London: Bradbury & Evans.
Dodd, S. (2016). Orientating disability studies to disablist austerity: Applying Fraser's insights. *Disability & Society*, *31*(2), 149–165.
Doel, M. A. (1996). A hundred thousand lines of flight: A machinic introduction to the nomad thought and scrumpled geography of Gilles Deleuze and Félix Guattari. *Environment and Planning D: Society and Space*, *14*(4), 421–439.
Doel, M. A. (1999). *Poststructuralist Geographies: The Diabolical Art of Spatial Science*. Washington, DC: Rowman & Littlefield.
Domina, T., Penner, A., & Penner, E. (2017). Categorical inequality: Schools as sorting machines. *Annual Review of Sociology*, *43*, 311–330.
Dorling, D. (2018). *Peak Inequality: Britain's Ticking Time Bomb*. Oxford: Policy Press.
Downey, D. B., & Condron, D. J. (2016). Fifty years since the Coleman Report: Rethinking the relationship between schools and inequality. *Sociology of Education*, *89*(3), 207–220.
Dyson, A., & Gallannaugh, F. (2008). Disproportionality in special needs education in England. *The Journal of Special Education*, *42*(1), 36.

Edmonds, D. (2015, November 5). The integrated school that could teach a divided town to live together. *The Guardian*. https://www.theguardian.com/news/2015/nov/05/integrated-school-waterford-academy-oldham

Edwards, C., & Maxwell, N. (2023). Disability, hostility and everyday geographies of un/safety. *Social and Cultural Geography*, *24*(1), 157–174.

Evans, B., Horton, J., & Skelton, T. (eds) (2016). *Play and Recreation, Health and Wellbeing*. Singapore: Springer.

Evans, R., Holt, L., & Skelton, T. (eds) (2017). *Methodological Approaches*. Singapore: Springer.

Facca, D., Gladstone, B., & Teachman, G. (2020). Working the limits of "giving voice" to children: A critical conceptual review. *International Journal of Qualitative Methods*, *19*, 1609406920933391.

Faucher-King, F., & Le Galès, P. (2010). *The New Labour Experiment: Change and Reform under Blair and Brown*. Stanford: Stanford University Press.

Featherstone, B., Gupta, A., Morris, K., & White, S. (2018). *Protecting Children: A Social Model*. Oxford: Policy Press.

Featherstone, B., Morris, K., & White, S. (2014). *Re-imagining Child Protection: Towards Humane Social Work with Families*. Oxford: Policy Press.

Fielding, M. (2004). Transformative approaches to student voice: Theoretical underpinnings, recalcitrant realities. *British Educational Research Journal*, *30*(2), 295–311.

Fine, B. (2002). *Social Capital Versus Social Theory*. Hove: Psychology Press.

Foucault, M. (1977). *Discipline and Punish*. New York: Random House.

Foucault, M. (1978). *The History of Sexuality, Vol. 1: An Introduction (Translated by Robert Hurley)*. New York: Pantheon.

Foucault, M. (1982). The subject and power. *Critical Inquiry*, *8*(4), 777–795.

Foucault, M. (1984a). *The History of Sexuality, Vol 3: The Care of the Self (Translated by Robert Hurley)*. New York: Vintage Books.

Foucault, M. (1984b). *The History of Sexuality, Vol. 2: The Use of Pleasure*. New York: Vintage Books.

Foucault, M. (2003). *Abnormal: Lectures at the Collège de France, 1974–1975*. London: Picador.

Franklin, A., & Brady, G. (2022). 'Voiceless' and 'vulnerable': Challenging how disabled children and young people were portrayed and treated during the COVID-19 pandemic in the UK and a call for action, in Turok-Squire, R. (ed) *Children's Experience, Participation, and Rights during COVID-19* (pp. 141–158). Berlin: Springer International Publishing.

Fraser, N. (1998). Heterosexism, misrecognition and capitalism. *New Left Review*, *228*, 140–149.

Fuller, C., & Geddes, M. (2008). Urban Governance under neoliberalism: New labour and the restructuring of state-space. *Antipode*, *40*(2), 252–282.

Gagen, E. A. (2015). Governing emotions: Citizenship, neuroscience and the education of youth. *Transactions of the Institute of British Geographers*, *40*(1), 140–152.

Gallacher, L. A. (2017). From milestones to wayfaring: Geographic metaphors and iconography of embodied growth and change in infancy and early childhood. *Geohumanities*. https://researchportal.northumbria.ac.uk/ws/portalfiles/portal/39899705/Revised_GeoHumanities_Submission_Body_for_PP0566.pdf

Gallagher, M. (2011). Sound, space and power in a primary school. *Social & Cultural Geography*, *12*(01), 47–61.

Garner, R. (2013, March 31). 'Gradgrind' Michael Gove's new curriculum is 'so boring that truancy will rise,' teachers warn. *The Independent*. www.independent.co.uk/news/education/education-news/gradgrind-michael-gove-s-new-curriculum-is-so-boring-that-truancy-will-rise-teachers-warn-8555591.html

Gibbons, S., & Telhaj, S. (2007). Are schools drifting apart? Intake stratification in English secondary schools. *Urban Studies*, *44*(7), 1281–1305.

Gibbons, S., & Telhaj, S. (2016). Peer effects: Evidence from secondary school transition in England. *Oxford Bulletin of Economics and Statistics*, *78*(4), 548–575.

Gibson-Graham, J. K. (2007). Beyond global vs. local: Economic politics outside the binary fame, in Herod, A. (ed) *Geographies of Power: Placing Scale* (2nd edition, pp. 25–60). Hoboken: Blackwell.

Gregson, N., & Rose, G. (2000). Taking Butler elsewhere: Performativities, spatialities and subjectivities. *Environment and Planning D: Society and Space*, *18*(4), 433–452.

Grenfell, M. J. (2004). *Pierre Bourdieu: Agent Provocateur*. London: Bloomsbury.

Grosz, E. (1994). *Volatile Bodies: Toward a Corporeal Feminism*. London/New York: Routledge.

Guthman, J., & Mansfield, B. (2013). The implications of environmental epigenetics A new direction for geographic inquiry on health, space, and nature-society relations. *Progress in Human Geography*, *37*(4), 486–504.

Hall, E., & Wilton, R. (2017). Towards a relational geography of disability. *Progress in Human Geography*, *41*(6), 727–744.

Hall, J. J. (2020). 'The word gay has been banned but people use it in the boys' toilets whenever you go in': Spatialising children's subjectivities in response to gender and sexualities education in English primary schools. *Social and Cultural Geography*, *21*(2), 162–185.

Hansen, N., & Philo, C. (2007). The normality of doing things differently: Bodies, spaces and disability geography. *Tijdschrift Voor Economische En Sociale Geografie*, *98*(4), 493–506.

Harker, C., Hörschelmann, K., & Skelton, T. (eds) (2017). *Conflict, Violence and Peace*. Singapore: Springer.

Harris, R., & Johnston, R. (2008). Primary schools, markets and choice: Studying polarization and the core catchment areas of schools. *Applied Spatial Analysis and Policy*, *1*(1), 59–84.

Harvey, D. (1989). *The Condition of Postmodernity*. Oxford: Blackwell.

Harvey, D. (2011). Roepke lecture in economic geography: Crises, geographic disruptions and the uneven development of political responses. *Economic Geography*, *87*(1), 1–22.

Harvey, D. (2020). *The Anti-Capitalist Chronicles*. London: Pluto Press.

Hastings, A., Bailey, N., Bramley, G., Gannon, M., & Watkins, D. (2015). *The Cost of The Cuts: The Impact on Local Government and Poorer Communities*. York: Joseph Rowntree Foundation.

Hemming, P. J. (2011). Meaningful encounters? Religion and social cohesion in the English primary school. *Social & Cultural Geography*, *12*(1), 63–81.

Heslop, P., & Glover, G. (2015). Mortality of people with intellectual disabilities in England: A comparison of data from existing sources. *Journal of Applied Research in Intellectual Disabilities*, *28*(5), 414–422.

Hetherington, L., Berlin, J., Bica, M., & Cannon, V. (2020). Spotlight on voices from the community: COVID-19 impacts on Gypsy, Roma and Traveller communities in England. *International Journal of Roma Studies*, *2*(2) 87–92.

Hevey, D. (1991). From self love to the picket line, in Lees, S. (ed) *Disability Arts and Culture Papers*. London: Shape Publications.

Hill, M., & Tisdall, K. (2014). *Children and Society*. London/New York: Routledge.

Hillier, M. (2022). Government has risked and lost "unacceptable" billions of taxpayers' money in its Covid response – and must account to the generations that will pay for it. *Committees, UK Parliament*. https://committees.parliament.uk/committee/127/public-accounts-committee/news/161243/government-has-risked-lost-unacceptable-billions-of-taxpayers-money-in-its-covid-response-and-must-account-to-the-generations-that-will-pay-for-it/

Hiriscau, I. E., Stingelin-Giles, N., Stadler, C., Schmeck, K., & Reiter-Theil, S. (2014). A right to confidentiality or a duty to disclose? Ethical guidance for conducting prevention research with children and adolescents. *European Child & Adolescent Psychiatry*, *23*, 409–416.

Holland, J., Reynolds, T., & Weller, S. (2007). Transitions, networks and communities: The significance of social capital in the lives of children and young people. *Journal of Youth Studies*, *10*(1), 97–116.

Holloway, S. L., & Valentine, G. (2000a). Spatiality and the new social studies of childhood. *Sociology*, *34*(4), 763–783.

Holloway, S. L., & Valentine, G. (eds) (2000b). *Children's Geographies: Playing, Living, Learning*. London/New York: Routledge.

Holloway, S. L., Holt, L., & Mills, S. (2019). Questions of agency: Capacity, subjectivity, spatiality and temporality. *Progress in Human Geography*, *43*(3), 458–477.

Holt, L. (2004). Children with mind – body differences: Performing disability in primary school classrooms. *Children's Geographies*, *2*(2), 219–236.

Holt, L. (2013). Exploring the emergence of the subject in power: Infant geographies. *Environment and Planning D: Society and Space*, *31*, 645–663.

Holt, L., & Bowlby, S. (2019). Gender, class, race, ethnicity and power in an elite girls' state school. *Geoforum*, *105*, 168–178.

Holt, L., & Evans, R. (2016). Geographies of children and young people. Methodological approaches: Introduction, in Evans, R., Holt, L., & Skelton, T. (eds) *Methodological Approaches* (pp. 1–13). Berlin: Springer.

Holt, L., & Murray, L. (2022). Children and COVID 19 in the UK. *Children's Geographies*, *20*(4), 487–494.

Holt, L., & Philo, C. (2023). Tiny human geographies: Babies and toddlers as non-representational and barely human life? *Children's Geographies*, *21*(5), 819–831.

Holt, L., Bowlby, S., & Lea, J. (2013). Emotions and the habitus: Young people with socio-emotional differences (re) producing social, emotional and cultural capital in family and leisure space-times. *Emotion, Space and Society*, *9*, 33–41.

Holt, L., Bowlby, S., & Lea, J. (2017). "Everyone knows me. . . . I sort of like move about": The friendships and encounters of young people with Special Educational Needs in different school settings. *Environment and Planning A*, *49*(6), 1361–1378.

Holt, L., Bowlby, S., & Lea, J. (2019a). Disability, special educational needs, class, capitals, and segregation in schools: A population geography perspective. *Population, Space and Place*, *25*(4), e2229.

Holt, L., Jeffries, J., Hall, E., & Power, A. (2019b). Geographies of co-production: Learning from inclusive research approaches at the margins. *Area*, *51*(3), 390–395.

Holt, L., Lea, J., & Bowlby, S. (2012). Special units for young people on the Autistic Spectrum in mainstream schools: Sites of normalisation, abnormalisation, inclusion, and exclusion. *Environment and Planning A*, *44*(9), 2191–2206.

Hopkins, P. E. (2013). *Young People, Place and Identity*. London: Routledge.

Hörschelmann, K., & Colls, R. (2009). Introduction, in Hörschelmann, K., & Colls, R. (eds) *Contested Bodies of Childhood and Youth* (pp. 1–21). Berlin: Springer.

Horton, J., & Kraftl, P. (2012). Clearing out a cupboard: Memory, materiality and transitions, in Jones, O., & Garde-Hansen, J. (eds) *Geography and Memory* (pp. 25–44). London: Palgrave Macmillan.

Horton, J., & Pimlott-Wilson, H. (eds) (2021). *Growing Up and Getting By: International Perspectives on Childhood and Youth in Hard Times*. Oxford: Policy Press.

Hughes, G. A. (2014). Syria and the perils of proxy warfare. *Small Wars and Insurgencies*, *25*(3), 522–538.

Hunt, E., Tuckett, S., Robinson, D., & Cruikshanks, R. (2022). *Covid-19 and Disadvantage Gaps in England 2021*. London: Education Policy Institute. October.

Hutchinson, J. (2021). *Identifying Pupils with Special Educational Needs and Disabilities*. London: Education Policy Institute. https://epi.org.uk/publications-and-research/identifying-send/

Huuki, T., & Renold, E. (2016). Crush: Mapping historical, material and affective force relations in young children's hetero-sexual playground play. *Discourse: Studies in the Cultural Politics of Education*, *37*(5), 754–769.

Ilie, S., Sutherland, A., & Vignoles, A. (2017). Revisiting free school meal eligibility as a proxy for pupil socio-economic deprivation. *British Educational Research Journal*, *43*(2), 253–274.

Intelligence and Security Committee of Parliament (2020). *Russia*. London: HMSO https://isc.independent.gov.uk/wp-content/uploads/2021/01/20200721_HC632_CCS001_CCS1019402408-001_ISC_Russia_Report_Web_Accessible.pdf

James, A., Jenks, C., & Prout, A. (1998). *Theorizing Childhood*. Oxford: Policy Press.

James, A. (2010). Interdisciplinarity–for better or worse. Children's Geographies, *8*(2), 215–216.

Jeffrey, C. (2012). Geographies of children and youth II: Global youth agency. *Progress in Human Geography*, *36*(2), 245–253.

Jeffrey, C. (2013). Geographies of children and youth III: Alchemists of the revolution? *Progress in Human Geography*, *37*(1), 145–152.

Jeffrey, C., & Dyson, J. (2021). Geographies of the future: Prefigurative politics. *Progress in Human Geography*, *45*(4), 641–658.

Jeffrey, C., & Dyson, J. (eds) (2008). *Telling Young Lives: Portraits of Global Youth*. Philadelphia: Temple University Press.

Jenkins, S. P., Micklewright, J., & Schnepf, S. V. (2008). Social segregation in secondary schools: How does England compare with other countries? *Oxford Review of Education*, *34*(1), 21–37.

Jones, O., & Garde-Hansen, J. (2012). Introduction, in Jones, O., & Garde-Hansen, J. (eds) *Geography and Memory: Explorations in Identity, Place and Becoming* (pp. 1–18). London: Palgrave Macmillan

Joseph Rowntree Foundation. (2022). *UK Poverty 2022: The Essential Guide to Understanding Poverty in the UK | JRF*. www.jrf.org.uk/report/uk-poverty-2022

Kallio, K. P., Mills, S., & Skelton, T. (eds) (2016). *Politics, Citizenship and Rights*. New York: Springer Reference.

Katz, C. (2001). On the grounds of globalization: A topography for feminist political engagement. *Signs*, *26*(4), 1213–1234.

Katz, C. (2002). Vagabond capitalism and the necessity of social reproduction. *Antipode*, *33*(4), 709–728. https://doi.org/10.1111/1467-8330.00207

Katz, C. (2004). *Growing up Global: Economic Restructuring and Children's Everyday Lives*. Minneapolis: University of Minnesota Press.

Katz, C. (2008). Bad elements: Katrina and the scoured landscape of social reproduction. *Gender, Place and Culture*, *15*(1), 15–29.

Katz, C. (2011). Accumulation, excess, childhood: Toward a countertopography of risk and waste. *Documents D'Anàlisi Geogràfica*, *57*(1), 47–60.

Katz, C. (2018). The angel of geography: Superman, Tiger Mother, aspiration management, and the child as waste. *Progress in Human Geography*, *42*(5), 723–740.

Kazda, L., Bell, K., Thomas, R., McGeechan, K., & Barratt, A. (2019). Evidence of potential overdiagnosis and overtreatment of attention deficit hyperactivity disorder (ADHD) in children and adolescents: Protocol for a scoping review. *BMJ Open*, *9*(11), e032327

Keightley, E., Pickering, M., (2012). *The Mnemonic Imagination*. Basingstoke: Palgrave Macmillan.

Khattab, N. (2009). Ethno-religious background as a determinant of educational and occupational attainment in Britain. *Sociology*, *43*(2), 304–322.

Kingsbury, P., & Pile, S. (2014). The unconscious, transference, dives, repetition and other things tied to Geography, in Kingsbury, P., & Pile, S. (eds) *Psychoanalytic Geographies* (pp. 1–49). London/New York: Routledge.

Kitchin, R. (1998). "Out of place", 'knowing one's place': Space, power and the exclusion of disabled people. *Disability & Society*, *13*(3), 343–356.

Kitchin, R., & Wilton, R. (2003). Disability activism and the politics of scale. *Canadian Geographer/Le Géographe Canadien*, *47*(2), 97–115.

Klein, M. F. (ed) (1991). *The Politics of Curriculum Decision-making: Issues in Centralizing the Curriculum*. New York: Suny Press.

Kohan, W. O., Olsson, L. M., & Aitken, S. C. (2015). 'Throwntogetherness': A travelling conversation on the politics of childhood, education and what a teacher does. *Revista Eletrônica de Educação*, *9*(3), 395–410.

Kraftl, P. (2013). Beyond 'voice', beyond 'agency', beyond 'politics'? Hybrid childhoods and some critical reflections on children's emotional geographies. *Emotion, Space and Society*, *9*, 13–23.

Kraftl, P. (2017). Memory and autoethnographic methodologies in children's geographies: Recalling past and present childhoods, in Evans, R., Holt, L., & Skelton, T. (eds) *Methodological Approaches* (pp. 23–46). Berlin: Springer.

Kraftl, P. (2020). *After Childhood: Re-Thinking Environment, Materiality and Media in Children's Lives*. London/New York: Routledge.

Kraftl, P., Horton, J., & Tucker, F. (2022). Children's geographies. *Oxford Bibliographies Online: Childhood Studies*. www.oxfordbibliographies.com/display/document/obo-9780199791231/obo-9780199791231-0080.xml

Kristeva, J. (1982). *Powers of Horror* (Vol. 98). Princeton: University Presses of California.

Kustatscher, M. (2017). The emotional geographies of belonging: Children's intersectional identities in primary school. *Children's Geographies*, *15*(1), 65–79.

Laurier, E., & Philo, C. (2006). Cold shoulders and napkins handed: Gestures of responsibility. *Transactions of the Institute of British Geographers*, *31*(2), 193–207.

Lea, J., Holt, L., & Bowlby, S. (2015). Reconstituting social, emotional and mental health difficulties: The use of restorative justice approaches in schools, in Blazek, M., & Kraftl, P. (eds) *Children's Emotions in Policy and Practice* (pp. 181–200). Basingstoke: Palgrave Macmillan.

Lens, V. (2019). Judging the other: The intersection of race, gender, and class in family court. *Family Court Review*, *57*(1), 72–87.

Lid, I. M. (2015). Vulnerability and disability: A citizenship perspective. *Disability & Society*, *30*(10), 1554–1567.

Lingard, B., & Thompson, G. (2017). Doing time in the sociology of education. *British Journal of Sociology of Education*, *38*(1), 1–12.

Lister, R. (2021). *Poverty*. Cambridge: Polity.

Lizardo, O. (2004). The cognitive origins of Bourdieu's habitus. *Journal for the Theory of Social Behaviour*, *34*(4), 375–401.

Lopes, J. J. M. (2014). Espaço desacostumado: A geografia das crianças e a geografia na educação infantil. *Revista Olh@res*, *2*(4), 301–334.

Lovell, T. (2000). Thinking feminism with and against Bourdieu. *Feminist Theory*, *1*(1), 11–32.

Lupton, D. (2013). Infant embodiment and interembodiment: A review of sociocultural perspectives. *Childhood*, *20*(1), 37–50.

Lupton, R., & Hayes, D. (2021). *Great Mistakes in Education Policy and How to Avoid Them in the Future*. Oxford: Policy Press.

Lupton, R., Hills, J., Steward, K., & Vizard, P. (2013). *Labour's Social Policy Record: Policy, Spending and Outcomes 1997–2010*. London: Trust for London. www.trustforlondon.org.uk/publications/labours-social-policy-record-policy-spending-and-outcomes-1997-2010/

Lupton, R., Thrupp, M., & Brown, C. (2010). Special educational needs: A contextualised perspective. *British Journal of Educational Studies*, *58*(3), 267–284.

Maguire, D. (2021). *Male, Failed, Jailed: Masculinities and "Revolving Door" Imprisonment in the UK*. Berlin: Springer.

Martino, W. (1999). "Cool boys", 'party animals", "squids" and "poofters": Interrogating the dynamics and politics of adolescent masculinities in school. *British Journal of Sociology of Education*, *20*(2), 239–263.

Massey, D. (1993). Power geometries and a progressive sense of place, in Bird, J., Curtis, B., Putnam, T., & Tickner, L. (eds) *Mapping the Futures: Local Cultures, Global Change* (pp. 75–85). London/New York: Routledge.
Massey, D. (2005). *For Space*. Thousand Oaks, CA: Sage.
Maxwell, S., Reynolds, K. J., Lee, E., Subasic, E., & Bromhead, D. (2017). The impact of school climate and school identification on academic achievement: Multilevel modeling with student and teacher data. *Frontiers in Psychology, 8*, 2069.
May, J., & Thrift, N. (2001). *Timespace: Geographies of Temporality*. London/New York: Routledge.
McGillicuddy, D., & Devine, D. (2020). 'You feel ashamed that you are not in the higher group' – Children's psychosocial response to ability grouping in primary school. *British Educational Research Journal, 46*(3), 553–573.
McIntyre, M., & Nast, H. J. (2011). Bio (necro) polis: Marx, surplus populations, and the spatial dialectics of reproduction and "race". *Antipode, 43*(5), 1465–1488.
McKee, M. (2020). Coronavirus shows how UK must act quickly before being shut out of Europe's health protection systems. *The BMJ, 369*. https://doi.org/10.1136/bmj.m400
McKittrick, K. (2011). On plantations, prisons, and a black sense of place. *Social & Cultural Geography, 12*(8), 947–963.
McNay, L. (1994). *Foucault: A Critical Introduction*. London: Continuum.
McNay, L. (2004). Agency and experience: Gender as a lived relation. *The Sociological Review, 52*, 173–190.
McRobbie, A. (2009). *The Aftermath of Feminism: Gender, Culture and Social Change*. Thousand Oaks, CA: Sage.
Millei, Z., & Imre, R. (2021). Banal and everyday nationalisms in children's mundane and institutional lives. *Children's Geographies, 19*(5), 505–513.
Mills, S. (2021). *Mapping the Moral Geographies of Education: Character, Citizenship and Values*. London/New York: Routledge.
Mills, S., & Waite, C. (2018). From big society to shared society? Geographies of social cohesion and encounter in the UK's National Citizen Service. *Geografiska Annaler: Series B, Human Geography, 100*(2), 131–148.
Mitchell, K. (2003). Educating the national citizen in neoliberal times: From the multicultural self to the strategic cosmopolitan. *Transactions of the Institute of British Geographers, 28*(4), 387–403.
Modood, T. (2004). Capitals, ethnic identity and educational qualifications. *Cultural Trends, 13*(2), 87–105.
Morris, J. (1991). *Pride against Prejudice: Transforming Attitudes to Disability*. London: The Women's Press.
Morrison, C. A. (2022). A personal geography of care and disability. *Social and Cultural Geography, 23*(7), 1041–1056.
Morris-Roberts, K. (2004). Girls' friendships, "distinctive individuality" and socio-spatial practices of (dis)identification. *Children's Geographies, 2*(2), 237–255.
Morrow, V. (1999). Conceptualising social capital in relation to the well-being of children and young people: A critical review. *The Sociological Review, 47*(4), 744–765.
Morrow, V. (2001). Young people's explanations and experiences of social exclusion: Retrieving Bourdieu's concept of social capital. *International Journal of Sociology and Social Policy, 21*(4/5/6).
Moser, S. C. (2016). Reflections on climate change communication research and practice in the second decade of the 21st century: What more is there to say?. *Wiley Interdisciplinary Reviews: Climate Change, 7*(3), 345–369.
Musson, S. (2010). The geography of UK Government Finance: Tax, spend and what lies in between, in Cole, N., & Jones, A. (eds) *The Economic Geography of the UK* (pp. 182–200). Thousand Oaks, CA: Sage.

Nairn, K., Kraftl, P., & Skelton, T. (eds) (2016). *Space, Place and Environment*. Singapore: Springer.
Nind, M. (2011). Participatory data analysis: A step too far?. *Qualitative Research, 11*(4), 349–363.
Nosworthy, C. (2014). *A Geography of Horse-riding: The Spacing of Affect, Emotion and (Dis) ability Identity through Horse-human Encounters*. Cambridge: Cambridge Scholars Publishing.
Nowotny, H. (1981). Women in public life in *Austria*, in Fuchs Epstein, C., & Laub Coser, R. (eds) *Access to Power* (pp. 147–156). London/New York: Routledge.
Olssen, M. (2004). The school as the microscope of conduction: Doing foucauldian research in education, in Marshall, J. D. (ed) *Poststructuralism, Philosophy, Pedagogy. Philosophy and Education*, vol. 12 (pp. 57–84). Berlin: Springer.
Olsson, L. M. (2009). Movement and Experimentation in Young Children's Learning: Deleuze and Guattari in early childhood education. London/New York: Routledge.
Pain, R., & Cahill, C. (2022). Critical political geographies of slow violence and resistance. *Environment and Planning C: Politics and Space, 40*(2), 359–372.
Parr, H. (1998). Mental health, ethnography and the body. *Area, 30*(1), 28–37.
Parr, H., & Butler, R. (1999). New geographies of illness, impairment and disability, in Butler, R., & Parr, H. (eds) *Mind and Body Spaces: Geographies of Illness, Impairment and Disability* (pp. 1–24). London/New York: Routledge.
Philo, C. (2007). A vitally human medical geography? Introducing Georges Canguilhem to geographers. *New Zealand Geographer, 63*(2), 82–96.
Philo, C. (2012). A 'new Foucault' with lively implications – Or 'the crawfish advances sideways'. *Transactions of the Institute of British Geographers, 37*(4), 496–514.
Philo, C., & Parr, H. (2000). Institutional geographies: Introductory remarks. *Geoforum, 31*(4), 513–521.
Philo, C., & Parr, H. (2003). Introducing psychoanalytic geographies. *Social and Cultural Geography, 4*(3), 283–93.
Pile, S. (1996). *The Body and the City: Psychoanalysis, Space, and Subjectivity*. London/New York: Routledge.
Pile, S. (2010). Emotions and affect in recent human geography. *Transactions of the Institute of British Geographers, 35*(1), 5–20. https://doi.org/10.1111/j.1475-5661.2009.00368.x
Pomerantz, S., & Raby, R. (2020). Bodies, hoodies, schools, and success: Post-human performativity and smart girlhood. *Gender and Education, 32*(8), 983–1000.
Prout, A. (2000). Childhood bodies: Construction, agency and hybridity, in Prout, A. (ed) *The Body, Childhood and Society* (pp. 1–18). London: Palgrave Macmillan.
Punch, S., & Tisdall, E. K. M. (2012). Exploring children and young people's relationships across majority and minority worlds. *Children's Geographies, 10*(3), 241–248
Putnam, R. (2000). Bowling Alone: The Collapse and Revival of American Community. London: Simon and Schuster.
Pykett, J. (2007). Making citizens governable? The Crick report as governmental technology. *Journal of Education Policy, 22*(3), 301–319.
Pykett, J., & Disney, T. (2016). Brain-targeted teaching and the biopolitical child, in Kallio, K. P., Mills, S., & Skelton, T. (eds) *Politics, Citizenship and Rights* (pp. 133–152). Singapore: Springer.
Raudenbush, S. W., & Eschmann, R. D. (2015). Does schooling increase or reduce social inequality. *Annual Review of Sociology, 41*, 443–470.
Reay, D. (2004a). 'It's all becoming a habitus': Beyond the habitual use of habitus in educational research. *British Journal of Sociology of Education, 25*(4), 431–444.
Reay, D. (2004b). Gendering Bourdieu's concepts of capitals? Emotional capital, women and social class. *The Sociological Review, 52*, 57–74.
Renold, E. (2005). *Girls, Boys, and Junior Sexualities: Exploring Children's Gender and Sexual Relations in the Primary School*. Hove: Psychology Press.

Richardson, H. (2021, October 5). School COVID absences rise two-thirds in fortnight. *BBC News*. www.bbc.co.uk/news/education-58805054

Riddell, S., & Weedon, E. (2016). Additional support needs policy in Scotland: Challenging or reinforcing social inequality? *Discourse: Studies in the Cultural Politics of Education, 37*(4), 496–512.

Ringrose, J. (2007). Successful girls? Complicating post-feminist, neoliberal discourses of educational achievement and gender equality. *Gender and Education, 19*(4), 471–489.

Rodda, P. C. (2022). Deep in Marvel's Closet: Heteronormativity and hidden LGBTQ+ narratives in the Marvel Cinematic Universe, in Carnes, N., & Goren, L. J. (eds) *The Politics of the Marvel Cinematic Universe* (pp. 266–277). Lawrence: University Press of Kansas.

Rose, N. (1990). *Governing the Soul: The Shaping of the Private Self*. London/New York: Routledge.

Ruddick, S. (2007). At the horizons of the subject: Neo-liberalism, neo-conservatism and the rights of the child part one: From 'knowing' fetus to 'confused' child. *Gender, Place and Culture, 14*(5), 513–527.

Ryan, F. (2020). *Crippled: Austerity and the Demonization of Disabled People*. London: Verso.

Sadler, S. (1999). *The Situationist City*. Cambridge: Massachusetts Institute of Technology press.

Saffer, J., Nolte, L., & Duffy, S. (2018). Living on a knife edge: The responses of people with physical health conditions to changes in disability benefits. *Disability & Society, 33*(10), 1555–1578.

Saldanha, A. (2010). Skin, affect, aggregation: Guattarian variations on Fanon. *Environment and Planning A, 42*(10), 2410–2427. https://doi.org/10.1068/A41288

Schaefer-McDaniel, N. J. (2004). Conceptualizing social capital among young people: Towards a new theory. *Children Youth and Environments, 14*(1), 153–172.

Schäfer, N., & Yarwood, R. (2008). Involving young people as researchers: Uncovering multiple power relations among youths. *Children's Geographies, 6*(2), 121–135.

Sclar, E. (2015). The political economics of investment Utopia: Public – private partnerships for urban infrastructure finance. *Journal of Economic Policy Reform, 18*(1), 1–15.

Shakespeare, T. (1994). Cultural representation of disabled people: Dustbins for disavowal? *Disability & Society, 9*(3), 283–299.

Sheller, M. (2017). From spatial turn to mobilities turn. *Current Sociology, 65*(4), 623–639.

Shildrick, M. (2005). Transgressing the law with Foucault and Derrida: Some reflections on anomalous embodiment. *Critical Quarterly, 47*(3), 30–46.

Shildrick, T., & MacDonald, R. (2013). Poverty talk: How people experiencing poverty deny their poverty and why they blame 'the poor.' *The Sociological Review, 61*(2), 285–303.

Sibieta, L. (2021). *School Spending in England: Trends Over Time and Future Outlook* (Briefing Note, 334). London: Institute for Fiscal Studies (IFS)

Simandan, D. (2016). Proximity, subjectivity, and space: Rethinking distance in human geography. *Geoforum, 75*, 249–252.

Sirin, S. R. (2005). Socioeconomic status and academic achievement: A meta-analytic review of research. *Review of Educational Research, 75*(3), 417–453.

Skelton, T. (2016). *Geographies of Children and Young People*. Singapore: Springer.

Smith, D. P., & Phillips, D. A. (2001). Socio-cultural representations of greentrified Pennine rurality. *Journal of Rural Studies, 17*(4), 457–469.

Smith, N. (2005). *The New Urban Frontier: Gentrification and the Revanchist City*. London/New York: Routledge.

Smørholm, S., & Simonsen, J. K. (2017). Children's drawings in ethnographic explorations: Analysis and interpretations, in Evans, R., Holt, L., & Skelton, T. (eds) *Methodological Approaches* (pp. 381–404). Berlin: Springer.

Staeheli, L. A. (2003). Cities and citizenship. *Urban Geography, 24*(2), 97–102.

Strand, S. (2016). Do some schools narrow the gap? Differential school effectiveness revisited. *Review of Education, 4*(2), 107–144.

Thomas, M. E. (2005). "I think it's just natural": The spatiality of racial segregation at a US high school. *Environment and Planning A*, *37*(7), 1233–1248.

Thomas, M. E. (2009). The identity politics of school life: Territoriality and the racial subjectivity of teen girls in LA. *Children's Geographies*, *7*(1), 7–19.

Thomas, M. E. (2010). Introduction: Psychoanalytic methodologies in geography. *Professional Geographer*, *62*(4), 478–482.

Thomas, M. E. (2011). *Multicultural Girlhood: Racism, Sexuality, and the Conflicted Spaces of American Education*. Philadelphia: Temple University Press.

Tolia-Kelly, D. P. (2006). Affect–an ethnocentric encounter? Exploring the 'universalist' imperative of emotional/affectual geographies. *Area*, *38*(2), 213–217.

Tomlinson, S. (2005). *Education in a Post-Welfare Society*. New York: McGraw-Hill Education.

Tyler, I. (2009). Against abjection. *Feminist Theory*, *10*(1), 77–98.

Tyler, I. (2013). *Revolting Subjects: Social Abjection and Resistance in Neoliberal Britain*. London: Bloomsbury Publishing.

Tyler, I. (2020). *Stigma: The Machinery of Inequality*. London: Bloomsbury Publishing.

UNICEF (2019). *An Unfair Start: Inequality in Children's Education in Rich Countries*. London: United Nations UK.

United Nations (1989). *Convention on the Rights of the Child*. Geneva: UN. www.hrweb.org/legal/child.html

United Nations News. (2021). Over 168 million children miss nearly a year of schooling, UNICEF says. *UN News*. https://news.un.org/en/story/2021/03/1086232

United Nations. (2015). *The Sustainable Development Agenda – United Nations Sustainable Development*. New York: UN. www.un.org/sustainabledevelopment/development-agenda/

Uprichard, E. (2008). Children as 'being and becomings': Children, childhood and temporality. *Children & Society*, *22*(4), 303–313.

Valentine, G. (2008). Living with difference: Reflections on geographies of encounter. *Progress in Human Geography*, *32*(3), 323–337.

Valentine, G., & Sporton, D. (2009). 'How other people see you, it's like nothing that's inside': The impact of processes of disidentification and disavowal on young people's subjectivities. *Sociology*, *43*(4), 735–751.

Valentine, G., Sporton, D., & Nielsen, K. B. (2009). Identities and belonging: A study of Somali refugee and asylum seekers living in the UK and Denmark. *Environment and Planning D: Society and Space*, *27*(2), 234–250

Vanderbeck, R. M., & Dunkley, C. M. (2004). Introduction: Geographies of exclusion, inclusion and belonging in young lives. *Children's Geographies*, *2*(2), 177–183.

Vincent, C. (2019). *Tea and the Queen? Fundamental British Values, Schools and Citizenship*. Oxford: Policy Press.

Wacquant, □L. (2009). *Punishing the Poor: The Neoliberal Government of Social Insecurity*[PDF]. Durham, NC: Duke University Press.

Waters, J. (2006). Geographies of cultural capital: Education, international migration and family strategies between Hong Kong and Canada. *Transactions of the Institute of British Geographers*, *31*(2), 179–192.

Watson, N. (2012). Theorising the lives of disabled children: How can disability theory help? *Children & Society*, *26*(3), 192–202.

Watson, N., McKie, L., Hughes, B., & Hopkins, D. (2004). (Inter) dependence, needs and care: The potential for disability and feminist theorists to develop an emancipatory model. *Sociology*, *38*(2), 331–350. https://doi.org/10.1177/0038038504040867

Webster, R., & Blatchford, P. (2015). Worlds apart? The nature and quality of the educational experiences of pupils with a statement for special educational needs in mainstream primary schools. *British Educational Research Journal*, *41*(2), 324–342.

West, C. (1989). *The American Evasion of Philosophy: A Genealogy of Pragmatism*. Berlin: Springer.

Wilson, A., & Urick, A. (2021). Cultural reproduction theory and schooling: The relationship between student capital and opportunity to learn. *American Journal of Education, 127*(2), 193–232.

Wilson, H. F. (2013). Collective life: Parents, playground encounters and the multicultural city. *Social & Cultural Geography, 14*(6), 625–648.

Wilson, H. F. (2014). Multicultural learning: Parent encounters with difference in a Birmingham primary school. *Transactions of the Institute of British Geographers, 39*(1), 102–114.

Wilson, H. F. (2017). On geography and encounter: Bodies, borders, and difference. *Progress in Human Geography, 41*(4), 451–471.

Woolf, V. (1972). *Moments of Being*. London: Random House.

Woolley, S. W. (2017). Contesting silence, claiming space: Gender and sexuality in the neoliberal public high school. *Gender and Education, 29*(1), 84–99.

Youdell, D. (2010). *School Trouble: Identity, Power and Politics in Education*. London: Routledge.

Zaharijević, A. (2016). In conversation with Judith Butler: Binds yet to be settled. *Filozofija i Društvo/Philosophy and Society, 71*(1), 105–114.

Appendices

Appendix One

Pseudonyms and characteristics of participants in the study

Young People

Aadesh, British Indian boy, unknown class background, with labels for Attention Deficit and Hyperactivity 'Disorder' and on the Autism Spectrum, rural high school, year eight.

Aashna, middle-class first-generation Indian migrant, urban selective girls' high school, year nine.

Adam, white British working-class boy on the Autism Spectrum, coastal special school, year 10.

Adi British Indian working-class boy, no Special Educational Need or Disability (SEND) label, coastal primary school, year five.

Aidan, white British middle-class boy on the Autism Spectrum, rural high school, year ten.

Alana, white British middle-class girl with moderate learning differences, rural special school, year eight or nine.

Alex, white British middle-class boy on the Autism Spectrum, rural high school, year ten.

Alfie, white British working-class boy with a visual impairment on the Autism Spectrum. Church Street, year five.

Ali, middle-class boy of Jordanian and Palestinian heritage from a middle-class family with a degenerative physical impairment who used a wheelchair, Rose Hill, year four.

Andrew, white middle-class British boy with specific learning differences, high school in a deprived coastal town, year nine.

Andy, white British boy from a poor adoptive family experiencing many difficulties on the Autism Spectrum and with moderate learning differences, rural special school, year five.

Annette, white British working-class girl with no SEND label, Church Street primary school, year four.

Anya, white British middle-class girl, selective girls' urban high school, year nine

Asha, white British middle-class girl with moderate learning differences, rural special school, year eight or nine.

Aya, British Asian girl with learning differences, unknown class background and a statement of SEND, Rose Hill, year four.

Ben, white British boy from a poor background, whose family had some involvement with social services and who had mild learning differences and physical impairments but no statement of SEND, year five.

Bevis, mixed heritage Black African and white British working-class boy, SEND label for specific learning difference, coastal primary school, year five.

Bryn, white British boy with moderate learning differences. We do not know if he had any additional labels or his class background, rural special school, year eight or nine.

Callum, white British working-class boy, from a poor background on the Autism Spectrum, coastal special school, year 10.

Clarence, white British working-class boy, no SEND label, coastal primary school, year five.

Conrad, working-class white British boy without SEND label, coastal high school, year seven.

Cora, British Chinese, middle-class girl, selective girls' urban high school, year nine.

Eden, middle-class, white British girl, without any labels of SEND, rural primary school, year five.

Ella, white British working-class girl on the Autism Spectrum and with an ADHD label, coastal special school, year ten.

Emilia, white British middle-class girl, focus group, selective urban high school, year nine.

Emma, white British middle-class girl, no labels of SEND, rural primary school, year five.

Erin, white British middle-class girl, urban selective girls' high school, year nine.

Erin, white British middle-class girl, selective girls' urban high school, year nine.

Estelle, white British girl with a label of Specific Learning Differences and from a working-class background, coastal primary school, year five.

Graham, white British boy, from a poor/socially excluded background and with specific learning differences, Church Street, year five.

Harriet, Black British upper-middle-class girl, urban selective girls' high school, year nine.

Holly, white British girl with visual impairments who used a wheelchair and whose parents were both out of formal employment, Seadale high school in a deprived coastal town, year nine.

Jacob, white British working-class boy with some mild learning differences whose family were experiencing difficulties, Rose Hill, year four.

Jasmin, British mixed heritage black Caribbean and white girl with labels of Social Emotional and Mental Health Difficulties (SEMHD), or with socio-emotional differences, from working-class background, coastal primary school, year five.

Joanna, white British working-class girl with Down's syndrome, Church Street, year five.

John, white British boy who has learning, social and emotional differences from a poor background, coastal primary school, year two.

Kaeya, British Indian, middle-class girl, selective girls' urban high school, year nine.

Karolina, white British middle-class girl, selective girls' urban high school, year nine.

Kasseem, middle-class British boy of Jordanian and Palestinian heritage, Rose Hill, with no SEND, year five, brother of Ali, who had a degenerative impairment and used a wheelchair, year four.

Kyle, white British working-class boy from a poor background, with specific learning differences and a label of Social Emotional and Mental Health 'Difficulties', coastal special school, year 10.

Lachlan, white British boy with labels of social, emotional and communication differences and from a poor family with 'difficulties' and a family history of learning differences, coastal primary school, year five.

Laura, white British girls middle-class girl, no labels of SEND, rural primary school, year five.

Leon, white British boy with a SEMHD label, moderate learning and medical differences, rural special school, year eight or nine.

Leon, white British working-class boy from a poor background unknown label, coastal special school, year 10.

Leroy, white British working-class boy with specific learning difficulties, coastal high school, year eight.

Liam, white British working-class boy, SEND label for a specific learning difference, coastal primary school, year five.

Lindsay, white British working-class girl with physical impairment who used a wheelchair, Rose Hill, year five.

Loretta, white British working-class girl on the Autism Spectrum, Church Street, year five.

Lorna, a white British girl, without SEND labels, although she was experiencing some difficulties with learning, Rose Hill, year four.

Lucia, white British girl, unknown class background, with moderate learning differences, rural special school, year eight or nine.

Lucy, white British working-class girl with mild learning differences but not a statement of SEND, Rose Hill, year five.

Mahal, British Bangladeshi working-class girl with labels of SEMHD, or with socio-emotional differences, coastal primary school, year five.

Mary, middle-class white British girl, no SEND label, rural primary school. All the girls discussed were middle-class white British girls without SEND, year five.

Megan, white British girl with complex learning and physical differences tied to brain damage at birth, poor background, coastal primary school, year two.

Nelson, white British working-class boy with learning and communication differences, Rose Hill, year five.

Nicola, white girl from a traveller background who now lived in a house, with some learning differences but no label, Church Street, year four.

Noel, white British boy from a poor family with many challenges and social services involvement, who often came to school dirty and with non-specific learning differences and some motor coordination differences, Church Street, year five.

Paavai, British Sri Lankan girl, unknown class origin, selective urban girls' high school, year nine.

Patty, white British middle-class girl, selective girls' urban high school, year nine

Rhana, middle-class girl from Saudi Arabia whose parents were working at the university and planned to return to Saudi Arabia, Rose Hill, year five.

Rosie, white British working-class girl living in foster care after being removed for abuse and neglect from her birth mother and with a progressive visual and hearing impairment, Church Street, year five.

Saabira, British Indian, middle-class girl, selective girls' urban high school, year nine.

Sabelle, British Turkish, middle-class girl, selective girls' urban high school, year nine

Sada, British Indian, middle-class girl, urban selective girls' high school, year nine.

Sally, white British middle-class girl, urban selective girls' high school, year nine.

Sam, white British working-class boy unknown label, coastal special school, year 10.

Samia, British Indian, middle-class girl, selective girls' urban high school, year nine.

Sharon, working-class white British girl with achondroplasia, who self-identified as a dwarf, with no other SEND but a 'statement', Church Street, year four.

Some characteristics of the schools are provided in Appendix One.

Summer, white British girl, unknown class background, no learning differences, Rose Hill, year four.

Tina, white British middle-class girl, selective girls' urban high school, year nine.

Violet, white British working-class young person with cystic fibrosis, high school in a deprived coastal town, year nine.

Adults

Mr Keegan, white British male class teacher, Church Street.

Mr Parker, headteacher, white British male, Rose Hill.

Mr Taylor, white British male class teacher, Rose Hill

Ms Robinson, white British female class teacher, Rose Hill.

Ms. Buttery, white British female teacher Church Street

Ms. Gregson, SENCO, white British female, Church Street.

Ms. Jessop, white British female LSA, Church Street

Ms. Massey, white British female class teacher, Church Street.

Ms. Miller, white British working-class LSA, Church Street.

Ms. Trim, headteacher, white British female Church Street.

Appendix Two

Overview of the schools

Church Street	Discussed in depth in Chapter 7. A primary school with high levels of SEND and free school meals in one of the most deprived neighbourhoods in the UK. Exclusively white, with a small number of gypsy-traveller children.
Rose Hill	Discussed in depth in Chapter 7. A primary school with relatively high levels of SEND and mean free school meals. In a suburb close to the city centre, with mixed population in terms of class and race/ethnicity. With some children of highly educated hyper-mobile parents working or studying at the University.
Seadale High School	A high school in a disinvested seaside town in the Southeast of England, with high levels of deprivation and a high proportion of young people with labels of SEND. Students almost exclusively of white British heritage.
Rural Special School	Largely white, 130 students, above-average Free School meals (30%). Age range 2–19. Complex learning differences.
Rural Primary School	Largely white, large primary school, with 235 students. Below-average Free School Meals (4%). Higher-than-average identified students with special educational needs.
Rural High School	Largely white, high school with 1,256 students. Low proportion of young people eligible for free school meals (4%). Lower-than-average proportion of young people with identified Special Educational Needs. Units for young people on the Autism Spectrum and with Social, Emotional and Mental Health differences (SEMHD).
Urban Special School	Racially and ethnically more mixed. Age range 5–11. Young people with SEMHD only. High proportion of children eligible for free school meals (47%).
Urban Primary School	Racially and ethnically more mixed. High proportion of students with SEND (33%), high proportion of free school meals (24%); 262 students aged 4–11.
Urban High School	Selective girl's high school situated in a large town. Prestigious with excellent results, and some students commuting large distances to attend. State school, not fee-paying. Racially and ethnically mixed. Below-average SEND (0.1%) and free school meals (0.7%). Around 700 students aged 11–18.
Coastal Special School	Largely white, 47 students. Age 11–16. All students with complex learning needs. Above-average free school meals (24%).
Coastal Primary School	Racially and ethnically more mixed. Above-average SEND (34%) and free school meals (24%). 210 students aged 4–11.
Coastal High School	Mostly white, around 960 students. Age 11–16. High percentage of SEND (27%) and free school meals (19%). Primary-style nurture classrooms to support young people transitioning to high school, especially those with SEND.

All figures of SEND are recognised SEND, being on the register of SEND or with individual interventions.

At the time of *this* research, the proportion of young people nationally who were eligible for free school meals was 15.9% in primary and 13.4% in high school, and 20.5% of students had some recognised Special Educational Need or Disability.

Index

abjection 106–107; defining 60; of disability 60–61, 103, 104–106, 145; resisting 115–116, 152
ableism: and school 10, 11, 53–54, 66, 104–105, 110, 118, 126, 127; in society 53, 105, 127; transformation of 50, 116–118, 152; young people reproducing 89–90, 102, 104
ADHD 81
adults: adult-child relationships 4, 14, 27–28, 38, 40, 66, 83, 129; assumptions of 91, 128–129; research with 33; *see also* teachers
agency: conceptualising 12, 13, 36, 38, 43–45, 127; critiquing 12, 38–39, 43–45; as expressed 110–111, 146; as overlooked 11–12, 15, 35, 39, 44; power of 12, 13, 14; structure/agency dualism 12–13, 47, 52; *see also* children and young people, as contextual bodies/subjectivities/agencies; transformation
Aitken, Stuart 39
aspirations 126, 135–136, 139, 146
attainment 9, 134; and SEND 145–146
Austerity 11, 26, 32, 145
Autism Spectrum (Disorders) 16–17, 90–91, 100–101, 105, 109–110, 129
autoethnography 5–8
Azpitarte, Fran 144

belonging 90, 102, 135
Benjamin, Jessica 48, 55, 57, 62
Benjamin, Walter 114–115
bodies: bodily competence 36, 46, 88–90, 91–92, 116–117, 123, body/society dualism 47; 'without organs' 70, 117–118; of young people 13, 36–37, 45–46, 47–48, 50, 51, 63, 73–74
Bondi, Liz 57, 62–63
Bourdieu, Pierre 41, 42, 46, 49, 52; intersection with Butler 46, 51–53; meritocracy 53; *see also* cultural capital; capital interconnections; economic capital; habitus; social capital
Bowlby, Sophie 26, 47
Brazil 10–11
bullying: experience of 78, 79–80, 101, 103–104, 106, 120–121; resistance to 88, 100, 115–116; vulnerability to 128, 152; *see also* SEND; teachers
Butler, Judith 46, 52, 53, 57, 61; exegesis of power 45, 52, 55; interdependence 42, 43, 48, 52; performativity 47, 48, 49; *The Psychic Life of Power (1977)* 59, 62; subjection 37, 45, 47, 48, 49, 51–52, 62, 66; *Undoing Gender (2004)* 62; *see also* Bourdieu; recognition

Cameron, David 51
capital interconnections 41–42, 49, 52, 86, 107, 110, 144, 146, 149
capitalism 20, 72, 151
Casey, Dame Louise 8, 149; The Casey Review into Opportunity and Integration 1
childhood 13, 36, 38, 47, 48–49, 124
children and young people: as becoming 37, 45–46, 70, 73–74; as contextual bodies/subjectivities/agencies 13, 36–37, 44–45, 48, 70, 88, 111, 114, 148, 151; as dynamic 36, 45–46, 48–49, 50, 70; as futurity 8, 15, 36, 50–51, 73–74, 114, 121, 124, 126–127, 133, 148–149; as nodes of intergenerational reproduction 13, 37, 74, 86, 110; as powerful 13, 24, 36, 37, 110, 111, 120, 148, 149; as social agents (*see* agency); *see also* Butler, Judith, subjection

children and youth geographies: theoretical grounding 38, 43, 64–65
Children's Commissioner 18, 25
Children's Geographies 38
child-wheelchair assemblage 70, 116–118, 151
citizens 11, 66, 121
co-location 14–15, 79, 150
Conservative Party: Conservative Liberal Coalition 15, 26, 32; governance 26, 49
contextual bodies/subjectivities/agencies 3, 5, 11, 13, 19–21, 33, 35–36, 44, 45, 48, 70, 75, 84, 89, 114, 124, 148–149, 151.
COVID-19 Pandemic 10, 71; and education 10–11, 17–18
cultural capital 8, 9, 11, 33, 41, 42, 86, 110, 111, 127–128, 144; *see also* capital interconnections
curriculum: hidden 66; limitations 39, 53, 65, 110, 129, 145; value beyond 15, 139; value of 33, 65–66, 123

dis/ability: ab/normalisation 45–46, 53–55, 60–61, 102, 104–105, 110, 127; costs 145–146; dependency 91, 116–117; disablism (*see* ableism); disablist slurs 104, 118; and disadvantage 9, 10, 88, 100, 104, 106–107, 127, 129, 145–146; individual tragedy model of 129; and transformation of 68, 70, 73, 116–118, 122; *see also* abjection; SEND

Economic and Social Research Council 26
economic capital 41, 42; *see also* capital interconnections
educational inequalities 8–9, 10, 11, 15, 31–32, 41, 42, 49, 53, 65, 110, 126, 132, 144–145, 151
educational policy 8–9, 11, 31–32, 51, 53; limits to 32, 36, 39, 45–46, 53, 126, 132, 139, 151; *see also* inclusion
education system (UK): formal 11, 108–109, 110, 149; independent 9, 42; mainstream schools 32–33, 128; and SEND 32–33, 50, 54, 100, 115, 123, 128; special education institution 33, 54, 100, 123, 128; state 9, 41
emotional capital 42–43, 76–77, 80, 81, 83–84, 96, 116, 151
empathy 15, 51, 61–63, 68–69, 124, 129; expressions of 75–77, 79, 116, 118–119, 120–121
encounter: geographies of 14, 15–16, 57–59, 60, 63–64, 66–67, 68–69; power of 15, 58–59, 70, 79, 121, 122–123, 124, 128–130, 150; in school 56–58, 64, 68, 70, 79, 121, 128–130; *see also* immersive geographies; transformation
Environmental and Physical Sciences Research Council 26
Eton 42

family: abusive 136; impact of 98, 136–137
feminist 12–13, 24, 35, 44, 64
Foucault, Michel 51, 53–54; ab/normalisation 45; self-regulation 45
free school meals 134, 141
friendships: as constrained 96–97, 98, 102, 126–129, 130–131, 146; and 'difference' 77, 87, 93, 97–98, 102, 107, 120–121, 122–123, 152; emotional (*see* emotional capital); and exclusivity 83–84, 88, 99–100; and formal education 108, 110, 149; gender (*see* gender); (dis)identification 35, 77–78, 88, 90, 91, 98, 99–100, 107, 115–116, 120, 123; importance of 75, 76–77, 80, 108, 110, 119–120, 148–149, 152; as overlooked 8, 12, 14, 17, 39–40, 83, 127–128; popularity 86, 87–88, 90, 99; power of 18, 37, 75–76, 91, 107, 110–111, 115, 116–117, 120, 124; precarious 77–80, 81–82, 82–84; race (*see* race); beyond school 86, 94, 96–98, 121, 124; and SEND 91, 100, 102, 116; as transformational 1, 88, 115–116, 118, 120–121, 123; transport 94, 96–97, 98; *see also* SEND

gender: and sociality 81–82, 87–89, 99, 149; *see also* LGBTQ+; SEND
Geographies of children and young people 38

habitus: conceptualising 37, 46–47, 48–49, 51, 61; external realities 47, 49; and school 49, 110, 114, 116; structuring 47; and young people 47, 48–49, 57, 70, 74, 77, 124, 151
Hegel, Georg Wilhelm Friedrich 48
homophobia 54, 93, 100

IMF 11
immersive geographies: conceptualising 2–5, 15–16, 19, 56–57, 58–59, 64–65, 68–70, 71, 114, 150–151; inequalities (*see* inequalities); school as site of 3, 15, 19, 56–57, 62, 64–65, 66–67, 114, 124–125, 150–151; transformation (*see* transformation); *see also* encounter; space-time relations

inclusion: inclusive education 8, 17, 32–33, 50–51, 54, 115, 123, 127–128, 129, 133, 141, 143–145, 152; inclusive society 14, 133
Indices of Multiple Deprivation 133
inequalities: challenging (*see* transformation); and intersectionality 24, 46, 49, 52, 105, 127, 144; reproducing 8–9, 41–42, 49, 51–52, 64, 74, 102–103, 105, 107, 146, 148–149; socio-economic (*see* social class); *see also* educational inequalities
infancy 44, 48
Institute of British Geographers 26
interior self 48, 50, 59–62, 63–64, 69, 70

Katz, Cindi: children as 'waste' 4, 27, 126–127, 146; counter-topographies 3, 16, 20, 24–25, 57, 72–73, 114, 124, 151–152; mimetic faculty 114–115; precious commodities 126
King, Martin Luther 1, 8, 120, 149

Labour Party 31, 32; education 31–32, 114; New Labour 31, 32
Lea, Jennifer 25, 26, 47, 82–83, 150
Learning Differences 104, 116, 118, 126–127, 129, 144–145; *see also* SEND
Learning Support 106, 128, 129
LGBTQ+ 54, 88
life course 13, 36, 47–48, 49, 139

Marxism 72, 151
Massey, Doreen 69, 71, 73; space and place 57, 67, 71; thrown togetherness 50, 57–58, 67; *see also* space-time relations
May, Therea 54
memory 5; collective 68, 69, 114, 124; mnemonic imagination 5
möbius topology 68, 69

National Curriculum 65
National Pupil Database 23
neoliberalism 10, 32, 41, 49, 53, 72, 106, 145, 146; education systems 8, 11, 39, 53
non-representational theory 35, 45, 59, 64
Northern Ireland 50

Ofsted 42, 132, 134, 140
othering 40, 45–46, 60, 93, 102–103, 105–106, 110; *see also* teachers

'peer effects' 12
play 4, 62, 73, 92–93, 115, 151
post-structuralist 12–13, 35, 44, 64
poverty 41, 94–95, 105, 106, 128, 133–134, 135–136, 137, 144, 146; and blame 41, 106, 137
privilege: class 9, 41, 42–43, 127; white 9; *see also* social class
Prout, Alan 38–39, 44
psychoanalytical geographies 43, 48, 50, 55, 59–62; *see also* Benjamin, Jessica; Hegel
Putnam, Robert 40–41

qualitative research 25–27, 29–30, 95; analysis 29–30; confidentiality 27–28, 29; disclosure 28–29; empowering research 3, 23–25, 148; ethics 27–29, 95; participatory research 24, 30; photo-voice 30; representativeness 26, 33; safeguarding 28

race/ethnicity 9; and racism 100, 120–121; and sociality 92–93, 97–98, 99–100, 102–103, 120–121
recognition 15, 43, 48, 52, 55, 57, 61–63, 68–69; expressions of 76–77, 84, 118, 120
risk society 137, 146
Royal Geographical Society 26

SATs 53, 134, 139, 140, 142
school: funding 32, 133–134, 138, 140, 142; institutionalisation 50, 53, 65, 67, 132; management 134, 138, 140, 142–143; materialities 88, 116, 119, 129–130; normalisation 11, 53–54, 65, 66, 91, 105, 127, 129, 149–150; power of 11, 15, 57, 58, 64, 65–66; resources 33, 128, 133–134, 138, 140–141, 142–143; socio-spatial context of 9, 33, 67, 70–71, 102, 126, 127–128, 132–133, 146; as unique space 14–15, 19, 65, 114, 126
'school climate' 9
SEMHD 105, 115
SENCO/SENDCO 134
SEND: and bullying 101, 103–104, 106, 108–109, 110, 152; and disadvantage 4, 10, 98, 103–105, 126–127, 128, 138, 142, 144; and exclusion 91–92, 94, 96, 98, 100, 101–102, 105, 129, 130–132, 152; friendships (*see* friendships); and inclusion 32, 115, 116–119, 122–123, 128; and marginalisation 89–91, 93, 94–95, 96–97, 105, 110, 128, 130–131; pupil referral unit 100, 110; resources for 123, 128, 130, 133–134, 138, 143–144, 145–146; from school 3, 26,

115, 119–120; and socio-economic disadvantage 9, 10, 94, 101, 106, 128, 144–146; *see also* ableism; dis/ability; education system; inclusion; social isolation
sexuality 54, 93
social capital 40–42, 75, 76, 84, 86, 96, 107, 110, 127, 148, 149; embodied 41; *see also* capital interconnections
social class 8–9, 14, 49, 52, 127, 144; *see also* poverty; privilege; SEND, socio-economic disadvantage
social isolation 80–81, 87, 89–90, 94–95, 96, 100–103, 105–106, 119–120, 130–132, 149, 152
social mobility 8, 31–32, 135–136, 137
social services 39, 136, 142
social tattooing 45
socio-emotional differences *see* learning differences
space-time relations 15, 56–57, 63–64, 66, 68–70, 124; circularity of time 16, 63–64, 68–69, 114, 121–122, 123, 124, 150; linearity of time 69; and school 68–70, 114, 126; time-space compression 71; *see also* immersive geographies
space/time/space/ time/space/time dialectic 16, 44, 56, 69, 150, 151
statements of Special Educational Needs 32, 133, 134, 138, 141

teachers: as bullying 103, 108–110, 131; as compassionate 107, 128, 132, 137; and intervention 83, 132; as othering 91, 109–110, 128–129, 131–132, 135–136, 137, 149–150; as overlooking social relationships 40, 83, 129, 132; power of 126–127, 128, 132, 140, 146, 149–150; and resource constraints 132, 134, 138, 139, 142
TimeSpace 15
transformation: with encounter 58, 62, 64, 66–67, 68–69, 124; expressions of 70, 88, 99, 116–118, 119–120, 122, 123; in friendships (*see* friendships); limits to 8, 15, 20, 58, 69, 124, 149, 150; potential for 1–3, 14, 37, 49–50, 57, 68–69, 73, 111, 114–115. 124, 150–151; in schools 49–50, 68, 70, 74, 114, 152; young people as agents of 13, 66, 73, 115, 116–118, 120, 122, 123, 148–149; *see also* ableism; immersive geographies

UK: education 9, 10–11, 32, 39, 65; politics 11, 20, 32, 49, 127; *see also* Austerity; education system
UN Convention on the Rights of the Child 36
UNICEF 10–11; Innocenti Report Card 9
UN Sustainable Development Goals 8
US 31, 39, 54

Waterhead Academy 50
wheelchair user 70, 90, 106, 116, 118; *see also* child-wheelchair assemblage
Woolf, Virginia 5